Verifiable Composition of Signature and Encryption

Laila El Aimani

Verifiable Composition of Signature and Encryption

A Comprehensive Study of the Design Paradigms

Laila El Aimani
École Nationale des Sciences Appliquées de Safi
Cadi Ayyad University
Safi, Morocco

ISBN 978-3-319-88551-3 ISBN 978-3-319-68112-2 (eBook)
https://doi.org/10.1007/978-3-319-68112-2

Softcover reprint of the hardcover 1st edition 2017

Printed on acid-free paper

This Springer imprint is published by Springer Nature
The registered company is Springer International Publishing AG
The registered company address is: Gewerbestrasse 11, 6330 Cham, Switzerland

To my family

Preface

Scope Cryptographic mechanisms that require the functionalities of both signature and encryption are becoming nowadays increasingly important.

Consider for example the case of interorganizational electronic documents; digital signatures on these documents are indispensable to resolve disputes as they ensure integrity and authenticity of the underlying messages; however, self-authentication of the signatures will make the messages vulnerable to industrial spy or extortionist. Therefore, it is imperative to control the signature verification by applying for instance an encryption layer that obscures the signatures and makes them opaque. Also, a secure email application requires signature and encryption simultaneously to guarantee confidentiality, integrity, and authenticity of the exchanged emails.

Verifiability is an important feature in those mechanisms. In fact, it can be applicable in filtering out spams in an email application; the spam filter should be able to check the validity of the encrypted email without decrypting it. Besides, the receiver that decrypts the email might be compelled, for instance to resolve later disputes, to prove that the sender is indeed the author of a given message. Likewise, the author of the opaque signature might need to prove its validity with respect to a given message, or to delegate this task to a third party.

This Book's Objectives This book attempts to give a thorough treatment of the celebrated compositions of signature and encryption that allow for good verifiability, i.e., possibility to efficiently prove properties about the encrypted data. The study is provided in the context of two cryptographic primitives: (1) designated confirmer signatures, an opaque signature which was introduced to control the proliferation of certified copies of documents, and (2) signcryption, a primitive that offers simultaneously and efficiently privacy and authenticity.

The choice of the case-study primitives is motivated by the need to have a representative of primitives that require both confidentiality and authenticity of the data, and a representative of opaque signatures which obfuscate the authenticity of the signed data while disclosing the latter. The hope is to be able to extend the present study to cover the plethora of cryptographic mechanisms that use both signature and encryption, and need good verifiability.

Instead of giving a compendium of results about the studied primitives, I take an instructive approach to first analyze and explain the shortcomings of the existing paradigms used to build the primitives, then proceed to the exposition of the efficient variants while giving the reader understanding and appreciation of the design methodology. Moreover, I endeavor to gradually supplement and reinforce the security model in which the primitives are being analyzed; the goal is to provide flexible design options according to the required security.

Audience The book is aimed at the following audiences.

- Researchers in cryptology/privacy. These readers will find a single-point reference book which gives a sound and rigorous treatment of the existing compositions of signature and encryption found in a great number of cryptographic and privacy-preserving mechanisms. Such a book can help this audience enter quickly into and master this vast area of study. It also presents an important literature survey material which can help them find further literature and consequently shape their own research topics.
- Graduate and PhD students beginning their research in cryptology and information security. These readers will find in this monograph a suitable cut-down set of many properly interwoven topics that form the basic pillars of modern cryptography; to name but a few: digital signatures, (tag-based) encryption, (non) interactive proofs, zero-knowledge, (meta) reductions.
- Security engineers in high-tech companies responsible for the design and development of cryptographic and privacy-preserving solutions. In fact, the book provides design principles and guidelines for certain cryptographic mechanisms in a pedagogical manner that allows to easily extend the study to further mechanisms. It constitutes then a suitable self-teaching text for this population in the area subject to the study.

Content The book is organized into four parts. There is a tight continuity from one part to the next to ensure a quick comprehension of the material. Thus, Part I (Chaps. 1 and 2) gives the necessary background in the theoretical foundations of modern cryptography, including the definition of the case-study primitives. Part II (Chaps. 3 and 4) and Part III (Chaps. 5 and 6) cover the existing compositions of signature and encryption, namely Sign_then_Encrypt (StE) and Commit_then_Encrypt_and_Sign (CtEaS) including the special instance Encrypt_then_Sign (EtS). Both parts start with a close scrutiny of the mentioned paradigms before putting forward the more efficient new analogs. Part IV (Chaps. 7–9) builds from the work developed in the previous parts to propound new paradigms that respond to stronger security requirements without compromising efficiency. Finally, we summarize in Chap. 10 the conclusions to be drawn from our study.

Acknowledgments

I developed most results presented in this book during my PhD and my postdoc at the University of Bonn and Technicolor respectively. It is a pleasure to thank my PhD supervisor Joachim von zur Gathen for his invaluable support and feedback during my studies. A special note of thanks go to Damien Vergnaud for making me discover cryptographic protocols and for his substantial help during the early stages of my PhD. I would also like to express my deep gratitude to my postdoc mentor Marc Joye for his generous support and countless suggestions to improve my results. My PhD reviewer Kenny Paterson deserves special mention for reading my thesis, a preliminary version of the presented results, in excruciating detail and giving me constructive comments that greatly improved the results and inspired me to derive new ones. I benefited from collaboration/correspondence with many researchers; I wish to thank all my colleagues and coauthors for precious discussions which were a great source of inspiration while writing this text. I am also grateful to Jorge Nakahara Jr. for encouraging me to turn my results into a book and for his excellent cooperation and availability throughout the edition process. Last but not least, I wish to express my profound gratitude to my family for constant understanding and endless support over the years. I am also indebted to my institute ESTS at Cadi Ayyad University for providing a nice working environment for completing this work.

Safi, Morocco Laila El Aimani
July 2017

Contents

Part I Background

1 Preliminaries 3
1.1 Cryptographic Primitives 3
1.1.1 Digital Signatures 3
1.1.2 Public-Key Encryption (PKE) 6
1.1.3 Key/Data Encapsulation Mechanisms 10
1.1.4 Tag-Based Encryption (TBE) 12
1.1.5 Commitment Schemes 14
1.2 Number-Theoretic Problems 16
1.2.1 Factoring-Related Problems 16
1.2.2 Discrete-Log-Related Problems 17
1.3 Reductionist Security 20
1.3.1 Cryptographic Reductions 20
1.3.2 Proof Models 22
1.3.3 Meta-reductions in Cryptography 23
1.4 Cryptographic Proof Systems 24
1.4.1 Interactive Proofs 24
1.4.2 Zero-Knowledge (ZK) 25
1.4.3 Σ Protocols 27
1.4.4 Non-interactive Proofs 28
References 29

2 Case-Study Primitives 31
2.1 Convertible Designated-Confirmer Signatures (CDCS) 31
2.1.1 Motivation 31
2.1.2 Syntax 32
2.1.3 Security Model for CDCS 33

2.2 Signcryption 39
2.2.1 Motivation and Challenges 39
2.2.2 Syntax 41
2.2.3 Security Model 42
References 43

Part II The "Sign_then_Encrypt" (StE) Paradigm

3 Analysis of StE 49
3.1 StE for Confirmer Signatures 49
3.1.1 The StE Paradigm 49
3.1.2 Other Variants 50
3.2 The Exact Unforgeability of StE Constructions 52
3.2.1 Roadmap for the Rest of the Chapter 53
3.3 A Breach in Invisibility with Homomorphic Encryption 53
3.4 Impossibility Results for Key-Preserving Reductions 54
3.4.1 Insufficiency of OW-CCA Secure Encryption 55
3.4.2 Insufficiency of NM-CPA Secure Encryption 56
3.4.3 Putting All Together 56
3.5 Extension to Arbitrary Reductions 57
3.6 Analysis of Damgård-Pedersen's Undeniable Signatures 60
3.7 Sufficiency of IND-PCA Secure Encryption 62
References 64

4 An Efficient Variant of StE 67
4.1 The New StE 67
4.1.1 Construction 67
4.1.2 Security Analysis 69
4.2 Practical Realizations 72
4.2.1 The Class $\mathbb{S}$ of Signatures 73
4.2.2 The Class $\mathbb{E}$ of Encryption Schemes 74
4.2.3 Confirmation/Denial Protocols 76
4.3 Further Enhancements 78
4.3.1 Reducing the Soundness Error 78
4.3.2 Online Non-transferability 79
4.4 Performance of the New StE 80
References 81

Part III The "Commit_then_Encrypt_and_Sign" (CtEaS) Paradigm

5 Analysis of CtEaS 85
5.1 CtEaS for Confirmer Signatures 85
5.2 The Exact Invisibility of CtEaS 87
5.2.1 Impossibility Results 87
5.2.2 Sufficiency of IND-PCA Secure Encryption 89
References 91

6 CtEtS: An Efficient Variant of CtEaS 93
6.1 Commit_then_Encrypt_then_Sign: CtEtS 93
6.1.1 The Construction 93
6.1.2 Security Analysis 94
6.1.3 Practical Instantiations 98
6.2 The "Encrypt_then_Sign" (EtS) Paradigm 99
6.2.1 Security Analysis 100
6.2.2 Confirmation/Denial Protocols 102
6.2.3 Selective Conversion 103
References 104

Part IV New Paradigms

7 EtStE: A New Paradigm for Verifiable Signcryption 107
7.1 Shortcomings of the Classical Paradigms 107
7.1.1 Review of the Classical Paradigms 107
7.1.2 Deficiencies of the New StE and CtEtS Paradigms 108
7.2 EtStE: A New Paradigm for Efficient Verifiable Signcryption 109
7.2.1 The Construction 110
7.2.2 Security Analysis 110
7.2.3 Practical Instantiations 112
References 113

8 Multi-User Security 115
8.1 Motivation and Definition 115
8.1.1 Formal Security Model 116
8.1.2 Extension to Multi-User Security 117
8.2 New Paradigms 118
8.2.1 Security Analysis 119
8.2.2 Performance 122
References 123

9 Insider Privacy 125
9.1 The CHK Transform 125
9.1.1 The CHK Transform for PKE 126
9.1.2 A CHK-Like Transform for TBE 127
9.2 New Paradigms with Insider Privacy 128
9.2.1 Constructions for Confirmer Signatures 130
9.2.2 Constructions for Verifiable Signcryption 132
9.2.3 Multi-User Security 133
9.2.4 Performance 134
References 136

10 Wrap-Up 137

Notational Index 139

Index 143

List of Definitions

1.1 Existential Unforgeability for Digital Signatures (EUF-CMA) 4
1.2 Strong Unforgeability for Digital Signatures (SEUF-CMA) 5
1.3 Strong Unforgeability for One-Time Signatures 5
1.4 Non-malleability for PKE (NM-CPA) 7
1.5 One-Wayness for PKE (OW-CCA) 7
1.6 Indistinguishability for PKE (IND-ATK) 8
1.7 Invisibility for PKE (INV-CPA) 9
1.8 Indistinguishability for KEM (IND-CPA) 10
1.9 Invisibility for DEMs (INV-OT) 11
1.10 Selective-Tag Indistinguishability for TBE (IND-st-wCCA) 13
1.11 Selective-Tag Indistinguishability for Tag-Based KEMs 13
1.12 Hiding in Commitment Schemes 14
1.13 Binding in Commitment Schemes 15
1.14 Injectivity in Commitment Schemes 15
1.15 Key-Preserving Reductions 21
1.16 Interactive Proofs (IP) 24
1.17 Zero-Knowledge 25

2.1 Completeness in CDCS 34
2.2 Soundness in CDCS 34
2.3 Non-transferability in CDCS 35
2.4 Unforgeability in CDCS (EUF-CMA) 37
2.5 Invisibility in CDCS (INV-CMA) 38
2.6 Unforgeability for Signcryption (EUF-CMA) 42
2.7 Indistinguishability for Signcryption (IND-CCA) 43

3.1 Homomorphic Encryption 53
3.2 Encryption with Non-malleable Key Generator 58
3.3 Indistinguishability in Damgård-Pedersen's Signatures 60

4.1 Strong Invisibility for Confirmer Signatures 70
4.2 Class $\mathbb{S}$ of Signatures 73

4.3 Class $\mathbb{E}$ of Encryption Schemes 74

6.1 Class $\mathbb{C}$ of Commitments 98

7.1 Strong Indistinguishability for Signcryption (SIND-CCA) 111

8.1 Multi-User Unforgeability for Confirmer Signatures 116
8.2 Multi-User Invisibility for Confirmer Signatures 117
8.3 Class $\mathbb{T}$ of Homomorphic Tag-Based Encryption 122

List of Figures

Fig. 1.1 Relations among security notions for PKE 7
Fig. 1.2 The hybrid encryption paradigm 12
Fig. 1.3 Pedersen's commitment scheme 16
Fig. 1.4 ElGamal's encryption scheme 21
Fig. 1.5 Example of a meta-reduction 23
Fig. 1.6 Σ protocol 28

Fig. 3.1 The StE paradigm 50

Fig. 4.1 The new StE paradigm 68
Fig. 4.2 Proof of knowledge of a preimage by function f 74
Fig. 4.3 Proof of knowledge of a decryption 75
Fig. 4.4 Confirmation/denial protocol for the new StE 77
Fig. 4.5 Proof of disjunctive knowledge 80

Fig. 5.1 The CtEaS paradigm 86

Fig. 6.1 The CtEtS paradigm 94
Fig. 6.2 Proof of knowledge of a decommitment 98
Fig. 6.3 Confirmation/denial protocol for CtEtS 99
Fig. 6.4 The EtS paradigm 100
Fig. 6.5 Denial protocol in EtS 103

Fig. 7.1 The EtStE paradigm for signcryption 110

Fig. 8.1 Multi-user StE for confirmer signatures 119
Fig. 8.2 Multi-user CtEtS for confirmer signatures 121
Fig. 8.3 Multi-user EtS for confirmer signatures 121
Fig. 8.4 Cash et al.'s tag-based encryption 122
Fig. 8.5 Proof of knowledge of a decryption in TBE 123

Fig. 9.1 The CHK transform 126
Fig. 9.2 The CHK-like transform for TBE 128
Fig. 9.3 StE with insider privacy 130

Fig. 9.4 CtEaS with insider privacy 131
Fig. 9.5 EtS with insider privacy 132
Fig. 9.6 EtStE with insider privacy 133
Fig. 9.7 Multi-user StE with insider privacy 134
Fig. 9.8 Multi-user CtEaS with insider privacy 135
Fig. 9.9 Multi-user EtS with insider privacy 135

Glossary

Cryptographic Systems

CDCS	Convertible Designated Confirmer Signature
Cryptosystem	Cryptographic System
DEM	Data Encapsulation Mechanism
FDH	Full-Domain Hash
KEM	Key Encapsulation Mechanism
OTS	One-Time Signature
PKE	Public Key Encryption
PKI	Public Key Infrastructure
TBE	Tag-Based Encryption

Security Notions

ANO	Anonymity
CCA	Chosen-Ciphertext Attack
CMA	Chosen-Message Attack
CPA	Chosen-Plaintext Attack
EUF	Existential Unforgeability
IND	Indistinguishability
INV	Invisibility
NM	Non-malleability
OW	One-Wayness
PCA	Plaintext-Checking Attack
SEUF	Strong Existential Unforgeability
SIND	Strong Indistinguishability
SINV	Strong Invisibility
st	Selective Tag
wCCA	Weak Chosen-Ciphertext Attack

Problems and Assumptions

CDH	Computational Diffie-Hellman
DDH	Decisional Diffie-Hellman
DL	Discrete Logarithm
GDH	Gap Diffie-Hellman

Proof Systems

CRS	Common Reference String
GS	Groth-Sahai
HVZK	Honest Verifier Zero-Knowledge
IP	Interactive Proof
NIZK	Non-interactive Zero-Knowledge
PoK	Proof of Knowledge
SpS	Special Soundness
WHPoK	Witness-Hiding Proof of Knowledge
ZK	Zero-Knowledge
ZKIP	Zero-Knowledge Interactive Proof
ZKP	Zero-Knowledge Proof
ZKPoK	Zero-Knowledge Proof of Knowledge

Miscellaneous

CtEaS	Commit_then_Encrypt_and_Sign
CtEtS	Commit_then_Encrypt_then_Sign
EtS	Encrypt_then_Sign
EtStE	Encrypt_then_Sign_then_Encrypt
NP	Non-polynomial
PPTM	Probabilistic Polynomial-Time Turing Machine
ROM	Random Oracle Model
StE	Sign_then_Encrypt

Part I
Background

Chapter 1
Preliminaries

Abstract This chapter serves an elementary-level introduction for the book. Section 1.1 introduces the most basic cryptographic primitives, namely digital signatures, public-key encryption including hybrid encryption (key/data encapsulation mechanisms) and tag-based encryption, and finally commitment schemes. The presentation of the primitives provides also the formal security notions that are needed later in our study. The following two sections consider an important notion of modern cryptography that is reductionist security: Sect. 1.2 recalls the frequently used intractable problems in cryptography, and Sect. 1.3 carries on the presentation of the basic tools used to gain confidence in cryptographic systems. Finally, Sect. 1.4 tackles an important cryptographic mechanism, needed in many real-life applications, that allows to conduct proofs without revealing more than the veracity of the proven statement.

1.1 Cryptographic Primitives

Notation Throughout the text, we will use a dot notation to refer the different components; for instance, $\Gamma.\texttt{encrypt}()$ denotes the encryption algorithm of public-key encryption scheme Γ, $\Sigma.pk$ the public key of signature scheme Σ, etc.

Besides, there is a `setup` algorithm, inherent to each of the presented primitives, that inputs a security parameter and generates the public parameters *param* of the scheme; although not always explicitly mentioned, *param* will serve as an input to all the algorithms/protocols that constitute the primitive in question.

Finally, the success probabilities of the experiments defining the different security notions that will be encountered, are taken over all the coin tosses of both the challenger and the adversary.

1.1.1 Digital Signatures

A digital signature scheme comprises the following algorithms:

$\texttt{keygen}(\mathbf{1}^\kappa)$ A probabilistic algorithm which inputs a security parameter κ and generates a pair of private and public keys.

L. El Aimani, *Verifiable Composition of Signature and Encryption*,
https://doi.org/10.1007/978-3-319-68112-2_1

$\mathtt{sign}_{sk}(\mathbf{m})$ A probabilistic algorithm which produces a signature on input a message m and a private key sk.

$\mathtt{verify}_{pk}(\boldsymbol{\sigma}, \mathbf{m})$ A deterministic algorithm which inputs a signature σ, a message m, and a public key pk, and outputs 1 if σ is a valid signature on m with respect to pk, and 0 otherwise.

We require that if (pk, sk) is a valid key pair, then

$$\forall m\colon \mathtt{verify}_{pk}(\mathtt{sign}_{sk}(m), m) = 1.$$

The standard security notion for a signature scheme is existential unforgeability under chosen-message attacks (EUF-CMA), which was introduced in Goldwasser et al. (1988). Informally, this notion refers to the hardness of, given a signing oracle, producing a valid pair of message and corresponding signature such that the message has not been queried to the signing oracle.

Definition 1.1 (Existential Unforgeability for Digital Signatures (EUF-CMA)**)** Let $\Sigma = (\mathtt{keygen}, \mathtt{sign}, \mathtt{verify})$ be a digital signature scheme, and let $\mathcal{A}$ be a PPTM. We consider the following random experiment, where κ is a security parameter.

Experiment $\mathbf{Exp}_{\Sigma,\mathcal{A}}^{\text{EUF-CMA}}(\kappa)$

1. $(pk, sk) \leftarrow \Sigma.\mathtt{keygen}(\kappa)$
2. $(m^\star, \sigma^\star) \leftarrow \mathcal{A}^{\mathfrak{S}}(pk)$
   $\mathfrak{S}\colon m \longmapsto \Sigma.\mathtt{sign}_{sk}(m)$
3. return 1 if and only if:
   - $\Sigma.\mathtt{verify}_{pk}[\sigma^\star, m^\star] = 1$,
   - and m was not queried to $\mathfrak{S}$.

We define the *advantage* of $\mathcal{A}$ *via*:

$$\mathbf{Adv}_{\Sigma,\mathcal{A}}^{\text{EUF-CMA}}(\kappa) = \Pr\left[\mathbf{Exp}_{\Sigma,\mathcal{A}}^{\text{EUF-CMA}}(\kappa) = 1\right],$$

where the probability is taken over the random tosses of both $\mathcal{A}$ and his challenger.

Given $(t, q_s) \in \mathbb{N}^2$ and $\varepsilon \in [0, 1]$, $\mathcal{A}$ is said to be a (t, ε, q_s)-EUF-CMA adversary against the scheme Σ if, running in time t and issuing q_s signing queries, $\mathcal{A}$ has $\mathbf{Adv}_{\Sigma,\mathcal{A}}^{\text{EUF-CMA}}(\kappa) \geq \varepsilon$. The scheme Σ is called (t, ε, q_s)-EUF-CMA secure if no (t, ε, q_s)-EUF-CMA adversary against it exists.

Remark 1.1 (Concrete Versus Asymptotic Security) The definition we provided above for digital signature quantifies clearly the advantage of an adversary when operating in a specific amount of time, and issuing a precise number of queries to the allowed oracles. There is another type of security which ignores those details and guarantees only a so-called asymptotic security, that is, all attacks vanish asymptotically if the attacker operates in time polynomial in the security parameter.

For example, a digital signature scheme $\Sigma(\kappa)$ with security parameter $\kappa \in \mathbb{N}$ is said to be EUF-CMA secure if, for any polynomial functions $t, q_s \colon \mathbb{N} \to \mathbb{N}$ and any non-negligible function $\varepsilon \colon \mathbb{N} \to [0, 1]$, it is $(t(\kappa), \varepsilon(\kappa), q_s(\kappa))$-EUF-CMA secure.

For the rest of this chapter, we provide only the concrete security for the various presented primitives; the asymptotic security can be easily derived.

Definition 1.2 (Strong Unforgeability for Digital Signatures (SEUF-CMA)**)** Consider the game in Definition 1.1. In case $\mathcal{A}$ is allowed to output a message already queried to $\mathfrak{S}$, yet not with a signature obtained from $\mathfrak{S}$, and has advantage ϵ when issuing q_s signing queries, then it is called a (t, ϵ, q_s)-SEUF-CMA adversary (S stands for "strongly"). Similarly, the scheme is said to be (t, ϵ, q_s)-SEUF-CMA secure if no (t, ϵ, q_s)-SEUF-CMA adversary against it exists.

One-Time Signatures One-time digital signature schemes can be used to sign, at most, one message; otherwise, signatures can be forged. A new public key is required for each message that is signed. One-time signature schemes have the advantage that signature generation and verification are very efficient. This is due to the fact that they rely on one-way functions without trapdoors. They are, similarly to normal digital signatures, defined by the key generation algorithm `keygen`, the signing algorithm `sign`, and the verification algorithm `verify`.

Definition 1.3 (Strong Unforgeability for One-Time Signatures) Let Σ = (`keygen`, `sign`, `verify`) be a one-time signature scheme, and let $\mathcal{A}$ be a PPTM. We consider the following random experiment, where κ is a security parameter.

Experiment $\mathbf{Exp}_{\Sigma,\mathcal{A}}^{\text{strong unforgeability}}(\kappa)$

1. $(pk, sk) \leftarrow \Sigma.\mathtt{keygen}(\kappa)$
2. $(m^\star, \sigma^\star) \leftarrow \mathcal{A}^{\mathfrak{S}}(pk)$
   $\mathfrak{S} \colon m \longmapsto \sigma = \Sigma.\mathtt{sign}_{sk}(m)$
3. return 1 if and only if the following properties are satisfied:
   - $\Sigma.\mathtt{verify}_{pk}[\sigma^\star, m^\star] = 1$
   - $(m, \sigma) \neq (m^\star, \sigma^\star)$ ($\mathfrak{S}$ is invoked on at most one message m)

We define the *advantage* of $\mathcal{A}$ *via*:

$$\mathbf{Adv}_{\Sigma,\mathcal{A}}^{\text{strong unforgeability}}(\kappa) = \Pr\left[\mathbf{Exp}_{\Sigma,\mathcal{A}}^{\text{strong unforgeability}}(\kappa) = 1\right].$$

Given $t \in \mathbb{N}$ and $\varepsilon \in [0, 1]$, $\mathcal{A}$ is said to be a (t, ε) strong unforgeability adversary against Σ if, running in time t, $\mathcal{A}$ has $\mathbf{Adv}_{\Sigma,\mathcal{A}}^{\text{strong unforgeability}}(\kappa) \geq \varepsilon$. The scheme Σ is called (t, ε) strongly unforgeable if no (t, ε) strong unforgeability adversary against it exists.

1.1.2 Public-Key Encryption (PKE)

A public-key encryption (PKE) scheme consists of the following algorithms.

$\texttt{keygen}(\mathbf{1}^\kappa)$ This is a probabilistic algorithm which generates a pair of private and public keys on input a security parameter κ.

$\texttt{encrypt}_{pk}(\mathbf{m})$ This is a probabilistic algorithm which takes as input a public key pk and a plaintext m and returns a ciphertext. Later in the text, we might use the notation $\texttt{encrypt}_{\{pk,coins\}}(m)$ to explicitly refer to the randomness *coins* used by the algorithm to generate the ciphertext.

$\texttt{decrypt}_{sk}(\mathbf{c})$ This is a deterministic algorithm which takes on input a private key sk and a ciphertext c, and returns the corresponding plaintext m or the symbol $\bot$.

We require that if (pk, sk) is a valid key pair, then

$$\forall m\colon \texttt{decrypt}_{sk}\left(\texttt{encrypt}_{pk}(m)\right) = m.$$

The typical *security goals* a PKE scheme should attain are: one-wayness (OW) that corresponds to the difficulty of inverting a ciphertext, indistinguishability (IND) that refers to the hardness of distinguishing ciphertexts based on the messages they encrypt, and finally non-malleability (NM) which corresponds to the hardness of deriving from a given ciphertext another ciphertext such that the underlying plaintexts are meaningfully related. Conversely, the typical *attack models* an adversary against an encryption scheme is allowed to are: Chosen-Plaintext Attack (CPA) where the adversary can encrypt any message of his choice—this scenario is unavoidable in a public-key framework—Plaintext-Checking Attack (PCA) in which the adversary is allowed to query an oracle on pairs (m, c) and gets answers whether c encrypts m or not, and finally Chosen-Ciphertext Attack (CCA) where the adversary is allowed to query a decryption oracle. Pairing the mentioned goals with these attack models yields nine *security notions*: GOAL-ATK for GOAL $\in$ {OW, IND, NM} and ATK $\in$ {CPA, PCA, CCA}. We provide in Fig. 1.1 the relations satisfied by these security notions; we refer to Bellare et al. (1998) for the details.

In the rest of this subsection, we define the notions for asymmetric encryption that we will encounter later in the text, namely NM-CPA, OW-CCA, and IND-ATK, for ATK $\in$ {CPA, PCA, CCA}.

The first notion that we consider is non-malleability under a chosen-plaintext attack (NM-CPA). It was introduced by Dolev et al. (1991), and is defined through a game between a challenger and an adversary $\mathcal{A}$. During this game, $\mathcal{A}$ can only encrypt messages of his choice, and at some point, he outputs to his challenger a distribution D from which the messages can be drawn. The challenger picks a message m from D, encrypts it in c and hands it to $\mathcal{A}$. $\mathcal{A}$ continues encrypting messages of his choice, and at the end of the game outputs a binary relation R and a ciphertext c'. $\mathcal{A}$ wins the game if the decryption of c' is related to m via the relation R, and the encryption scheme is proclaimed non-malleable if the advantage of $\mathcal{A}$ in this game is negligible (in the security parameter).

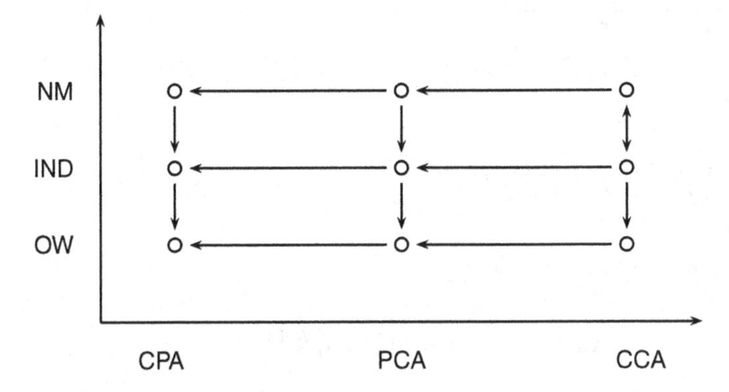

Fig. 1.1 Relations among security notions for PKE

Definition 1.4 (Non-malleability for PKE (NM-CPA)**)** Let Γ = (keygen, encrypt, decrypt) be a public-key encryption scheme, and let $\mathcal{A}$ be a PPTM. We consider the following random experiment, where κ is a security parameter:

Experiment $\mathbf{Exp}_{\Gamma,\mathcal{A}}^{\text{NM-CPA}}(\kappa)$

1. $(pk, sk) \leftarrow \Gamma.\mathtt{keygen}(\kappa)$
2. $D \leftarrow \mathcal{A}(pk)$
3. $m \overset{R}{\leftarrow} D$; $c \leftarrow \Gamma.\mathtt{encrypt}_{pk}(m)$
4. $(c', R) \leftarrow \mathcal{A}(pk, c)$
5. return (D, R, c, c')

We define the *advantage* of $\mathcal{A}$ *via*:

$$\mathbf{Adv}_{\Gamma,\mathcal{A}}^{\text{NM-CPA}}(\kappa) = \Pr[R(m, m')] - \Pr[R(m^\star, m')],$$

where $\mathbf{Exp}_{\Gamma,\mathcal{A}}^{\text{NM-CPA}}(\kappa) = (D, R, c, c')$, $m' = \Gamma.\mathtt{decrypt}_{sk}(c')$, and $m^\star \overset{R}{\leftarrow} D$. Given $t \in \mathbb{N}$ and $\varepsilon \in [0, 1]$, $\mathcal{A}$ is said to be a (t, ε)-NM-CPA adversary against Γ if, running in time t, $\mathcal{A}$ has $\mathbf{Adv}_{\Gamma,\mathcal{A}}^{\text{NM-CPA}}(\kappa) \geq \varepsilon$. The scheme Γ is called (t, ε)-NM-CPA secure if no (t, ε)-NM-CPA adversary against it exists.

The next notion is one-wayness under a chosen-ciphertext attack OW-CCA. one-wayness is the oldest and most natural notion public-key encryption should satisfy. It was introduced in the seminal work of Diffie and Hellman (1976) to denote the hardness of recovering plaintexts from their corresponding ciphertexts in a given encryption scheme. One-wayness under a chosen-ciphertext attack refers to the hardness of inverting ciphertexts even in presence of a decryption oracle the adversary can query on any ciphertext except of course on the challenge.

Definition 1.5 (One-Wayness for PKE (OW-CCA)**)** Let Γ = (keygen, encrypt, decrypt) be a public-key encryption scheme with message space $\mathcal{M}$, and let $\mathcal{A}$ be a PPTM. We consider the following random experiment, where κ is a security parameter.

Experiment $\mathbf{Exp}_{\Gamma,\mathcal{A}}^{\mathsf{OW\text{-}CCA}}(\kappa)$

1. $(pk, sk) \leftarrow \Gamma.\mathtt{keygen}(\kappa)$
2. $\mathcal{I} \leftarrow \mathcal{A}^{\mathtt{decrypt}(sk,.)}(pk)$
3. $m^\star \stackrel{R}{\leftarrow} \mathcal{M}$; $c^\star \leftarrow \Gamma.\mathtt{encrypt}_{pk}(m^\star)$
4. $\tilde{m} \leftarrow \mathcal{A}^{\mathtt{decrypt}\neg(c^\star)(sk,.)}(pk, c^\star)$
5. return $(\tilde{m} = m^\star)$

We define the *advantage* of $\mathcal{A}$ *via*:

$$\mathbf{Adv}_{\Gamma,\mathcal{A}}^{\mathsf{OW\text{-}CCA}}(\kappa) = \Pr\left[\mathbf{Exp}_{\Gamma,\mathcal{A}}^{\mathsf{OW\text{-}CCA}}(\kappa) = 1\right].$$

Given $(t, q_d) \in \mathbb{N}^2$ and $\varepsilon \in [0, 1]$, $\mathcal{A}$ is called a (t, ε, q_d)-OW-CCA adversary against Γ if, running in time t and issuing q_d decryption queries, $\mathcal{A}$ has $\mathbf{Adv}_{\Gamma,\mathcal{A}}^{\mathsf{OW\text{-}CCA}}(\kappa) \geq \varepsilon$. The scheme Γ is said to be (t, ε, q_d)-OW-CCA secure if no (t, ε, q_d)-OW-CCA adversary against it exists.

The last security notion we consider is indistinguishability or semantic security. It was introduced by Goldwasser and Micali (1984) and informally denotes the hardness of distinguishing ciphertexts based on their underlying messages. The formal definition of this notion is again through a game between an adversary $\mathcal{A}$ and a challenger. The game runs in three phases; in the first phase, $\mathcal{A}$ has access to the oracles allowed by the given attack model, and eventually outputs two messages $m_0^\star, m_1^\star$ from the message space considered by the given encryption scheme. In the second or challenge phase, the challenger picks uniformly at random one of the messages, encrypts it and gives the result to $\mathcal{A}$. In the last phase, $\mathcal{A}$ continues querying the oracles he had access to in the first phase, which now reject queries made w.r.t. the challenge ciphertext. At the end of the last phase, $\mathcal{A}$ outputs his guess for the message underlying the challenge, and is considered successful if the guess is correct.

Definition 1.6 (Indistinguishability for PKE (IND-ATK)**)** Let $\Gamma = (\mathtt{keygen}, \mathtt{encrypt}, \mathtt{decrypt})$ be a public-key encryption scheme, and let $\mathcal{A}$ be a PPTM. We consider the following random experiment, for $b \stackrel{R}{\leftarrow} \{0, 1\}$, where κ is a security parameter.

Experiment $\mathbf{Exp}_{\Gamma,\mathcal{A}}^{\mathsf{IND\text{-}ATK\text{-}}b}(\kappa)$

1. $(pk, sk) \leftarrow \Gamma.\mathtt{keygen}(\kappa)$
2. $(m_0^\star, m_1^\star, \mathcal{I}) \leftarrow \mathcal{A}^{\mathfrak{O}}(\mathsf{find}, pk)$
   - if ATK = CPA then $\mathfrak{O}$ = empty
   - if ATK = PCA then $\mathfrak{O}$: $(m, c) \longmapsto m \stackrel{?}{=} \Gamma.\mathtt{decrypt}_{sk}(c)$
   - if ATK = CCA then $\mathfrak{O}$: $c \longmapsto \Gamma.\mathtt{decrypt}_{sk}(c)$
3. $c^\star \leftarrow \Gamma.\mathtt{encrypt}_{pk}(m_b^\star)$

4. $d \leftarrow \mathcal{A}^{\mathfrak{O}}(\mathsf{guess}, \mathcal{I}, c^\star)$
   - if ATK = CPA then $\mathfrak{O}$ = empty
   - if ATK = PCA then $\mathfrak{O}: (m, c)(\neq (m_i^\star, c^\star), i = 0, 1) \longmapsto m \stackrel{?}{=} \Gamma.\mathtt{decrypt}_{sk}(c)$
   - if ATK = CCA then $\mathfrak{O}: c(\neq c^\star) \longmapsto \Gamma.\mathtt{decrypt}_{sk}(c)$
5. return (d)

We define the *advantage* of $\mathcal{A}$ *via*:

$$\mathbf{Adv}_{\Gamma,\mathcal{A}}^{\mathsf{IND\text{-}ATK}}(\kappa) = \left|\Pr\left[\mathbf{Exp}_{\Gamma,\mathcal{A}}^{\mathsf{IND\text{-}ATK}\text{-}b}(\kappa) = b\right] - \frac{1}{2}\right|.$$

Given $(t, q) \in \mathbb{N}^2$ and $\varepsilon \in [0, 1]$, $\mathcal{A}$ is called a (t, ε, q)-IND-ATK adversary against Γ if, running in time t and issuing q queries (to the allowed oracle), $\mathcal{A}$ has advantage greater than ε in the above experiment. The scheme Γ is said to be (t, ε, q)-IND-ATK secure if no (t, ε, q)-IND-ATK adversary against it exists.

Later in the text, we will further need the INV-CPA notion, i.e. invisibility under a chosen-plaintext attack, which denotes the difficulty to distinguish ciphertexts on an adversarially chosen message from random elements in the ciphertext space. Actually, this notion is the public-key variant of the INV-OT notion that we define later for Data Encapsulation Mechanisms, i.e. DEMs.

Definition 1.7 (Invisibility for PKE (INV-CPA)) Let $\Gamma = (\mathtt{keygen}, \mathtt{encrypt}, \mathtt{decrypt})$ be a public-key encryption scheme with ciphertext space C (the ciphertext space depends solely on the security parameter). Let further $\mathcal{A}$ be a PPTM. We consider the following random experiment, for $b \stackrel{R}{\leftarrow} \{0, 1\}$, where κ is a security parameter.

Experiment $\mathbf{Exp}_{\Gamma,\mathcal{A}}^{\mathsf{INV\text{-}CPA}\text{-}b}(1^\kappa)$

1. $(sk, pk) \leftarrow \Gamma.\mathtt{keygen}(1^\kappa)$
2. $(m^\star, \mathcal{I}) \leftarrow \mathcal{A}(1^\kappa)$
3. $e_1^\star \leftarrow \Gamma.\mathtt{encrypt}_{pk}(m^\star)$; $e_0^\star \stackrel{R}{\leftarrow} \mathsf{C}$
4. $d \leftarrow \mathcal{A}(\mathcal{I}, e_b^\star)$
5. return (d)

We define the *advantage* of $\mathcal{A}$ *via*:

$$\mathbf{Adv}_{\Gamma,\mathcal{A}}^{\mathsf{INV\text{-}CPA}}(1^\kappa) = \left|\Pr\left[\mathbf{Exp}_{\Gamma,\mathcal{A}}^{\mathsf{INV\text{-}CPA}\text{-}b}(1^\kappa) = b\right] - \frac{1}{2}\right|.$$

Γ is said to be (t, ϵ)-INV-CPA secure if the advantage $\mathbf{Adv}_{\Gamma,\mathcal{A}}^{\mathsf{INV\text{-}CPA}}(\kappa)$ of any adversary $\mathcal{A}$, operating in time t and running the experiment above, is no greater than ϵ.

Remark 1.2 (Invisibility vs Indistinguishability in PKE) It is not hard to see that breaking invisibility of an encryption scheme is easier than breaking its indistinguishability. In this sense, invisibility is a stronger assumption than semantic security in public-key encryption. To illustrate this fact, we can imagine an encryption scheme than appends the public key to the ciphertext; such a scheme might be indistinguishable, however it is clearly not invisible as one can easily distinguish encryptions on adversarially chosen messages from random elements in the ciphertext space (the ciphertext space is determined only by the security parameter). Actually, invisibility implies both indistinguishability and key-privacy or anonymity, that is, the difficulty to distinguish ciphertexts based on the keys under which they are created.

1.1.3 Key/Data Encapsulation Mechanisms

1.1.3.1 Key Encapsulation Mechanisms (KEM)

A KEM comprises three algorithms:

$\mathtt{keygen}(\mathbf{1}^\kappa)$ The key generation algorithm which probabilistically generates a pair of public and private keys on input a security parameter κ.

$\mathtt{encap}_{pk}()$ The encapsulation algorithm which inputs the public key pk and probabilistically generates a *session key* denoted k and its *encapsulation* c.

$\mathtt{decap}_{sk}(\mathbf{c})$ The decapsulation algorithm $\mathtt{decap}$ which inputs the private key sk and the element c and computes the decapsulation k of c, or returns $\perp$ if c is invalid.

We require for all valid pairs (pk, sk) the following:

$$\text{if } (c, k) = \mathtt{encap}_{pk}() \text{ then } k = \mathtt{decap}_{sk}(c).$$

Definition 1.8 (Indistinguishability for KEM (IND-CPA)) Let κ be a security parameter. Consider the following game conducted by a challenger and an adversary $\mathcal{A}$ against a KEM $\mathcal{K}$, for $b \xleftarrow{R} \{0, 1\}$ (K denotes the key space in the experiment below).

Experiment $\mathbf{Exp}_{\mathcal{K},\mathcal{A}}^{\mathsf{IND\text{-}CPA}\text{-}b}(1^\kappa)$

1. $(pk, sk) \leftarrow \mathcal{K}.\mathtt{keygen}(1^\kappa)$
2. $\mathcal{I} \leftarrow \mathcal{A}(pk)$
3. $(c^\star, k_1^\star) \leftarrow \mathcal{K}.\mathsf{encap}_{pk}()$; $k_0^\star \xleftarrow{R} \mathsf{K}$
4. $d \leftarrow \mathcal{A}(\mathcal{I}, c^\star, k_b^\star)$
5. return (d)

A KEM is (t, ϵ)-IND-CPA secure if the advantage, defined by

$$\mathbf{Adv}_{\mathcal{K},\mathcal{A}}^{\mathsf{IND\text{-}CPA}}(\kappa) = \left| \Pr\left[\mathbf{Exp}_{\mathcal{K},\mathcal{A}}^{\mathsf{IND\text{-}CPA}\text{-}b}(\kappa) = b \right] - \frac{1}{2} \right|,$$

of any adversary $\mathcal{A}$, operating in time t, in the above game, is no greater than ϵ.

1.1.3.2 Data Encapsulation Mechanisms (DEM)

A DEM is a secret-key encryption scheme given by the same algorithms forming a public-key encryption scheme that are:

`keygen`$(\mathbf{1}^\kappa)$ The key generation algorithm which produces uniformly distributed keys k on input a given security parameter.

`encrypt`$_{\mathbf{k}}(\mathbf{m})$ The encryption algorithm `encrypt` which inputs a key k and a message m and produces a ciphertext c.

`decrypt`$_{\mathbf{k}}(\mathbf{c})$ The decryption algorithm which decrypts ciphertext c using the same key k (used for encryption) to get back the message m or the special rejection symbol $\perp$.

We require that if k is a valid key, then

$$\forall m\colon \mathtt{decrypt}_k(\mathtt{encrypt}_k(m)) = m.$$

We define in the following invisibility under a one-time attack (INV-OT) for DEMs; this notion informally denotes the difficulty to distinguish encryption of an adversarially chosen message from a random ciphertext.

Definition 1.9 (Invisibility for DEMs (INV-OT)) Let κ be a security parameter. Consider the following game conducted by a challenger and an adversary $\mathcal{A}$ against a DEM $\mathcal{D}$, for $b \overset{R}{\leftarrow} \{0,1\}$ (C denotes the ciphertext space in the experiment below).

Experiment $\mathbf{Exp}_{\mathcal{D},\mathcal{A}}^{\mathsf{INV\text{-}OT}\text{-}b}(1^\kappa)$

1. $k \leftarrow \mathcal{D}.\mathtt{keygen}(1^\kappa)$
2. $(m^\star, \mathcal{I}) \leftarrow \mathcal{A}(1^\kappa)$
3. $e_1^\star \leftarrow \mathcal{D}.\mathtt{encrypt}_k(m^\star)$; $e_0^\star \overset{R}{\leftarrow} \mathsf{C}$
4. $d \leftarrow \mathcal{A}(\mathcal{I}, e_b^\star)$
5. return (d)

We define the *advantage* of $\mathcal{A}$ *via*:

$$\mathbf{Adv}_{\mathcal{D},\mathcal{A}}^{\mathsf{INV\text{-}OT}}(1^\kappa) = \left|\Pr\left[\mathbf{Exp}_{\mathcal{D},\mathcal{A}}^{\mathsf{INV\text{-}OT}\text{-}b}(1^\kappa) = b\right] - \frac{1}{2}\right|.$$

A DEM is said to be (t,ϵ)-INV-OT secure if the advantage $\mathbf{Adv}_{\mathcal{D},\mathcal{A}}^{\mathsf{INV\text{-}OT}}(\kappa)$ of any adversary $\mathcal{A}$, operating in time t and running the experiment above, is no greater than ϵ.

A DEM with injective encryption is a DEM where, for a every fixed key, the encryption algorithm `encrypt`, seen a function of the message, is injective, that is, for a fixed key, for every message m, there exists only one valid ciphertext that decrypts to m. Note that such DEMs exist, e.g. the one-time pad, and proffer interesting security properties, e.g. INV-OT.

$$\begin{array}{ll}
(\mathcal{K},\mathcal{D}).\mathsf{keygen}(1^\kappa) & : (sk,pk) \leftarrow \mathcal{K}.\mathsf{keygen}(1^\kappa) \\
 & \quad \mathsf{return}\ (sk,pk) \\
(\mathcal{K},\mathcal{D}).\mathsf{encrypt}_{pk}(m) & : (k,c) \leftarrow \mathcal{K}.\mathsf{encap}_{pk}()\ ;\ e \leftarrow \mathcal{D}.\mathsf{encrypt}_k(m) \\
 & \quad \mathsf{return}\ (c,e) \\
(\mathcal{K},\mathcal{D}).\mathsf{decrypt}_{sk}(c,e) & : k \leftarrow \mathcal{K}.\mathsf{decap}_{sk}(c)\ ;\ \tilde{m} \leftarrow \mathcal{D}.\mathsf{decrypt}_k(e) \\
 & \quad \mathsf{return}\ (\tilde{m})
\end{array}$$

Fig. 1.2 The hybrid encryption paradigm

1.1.3.3 The Hybrid Encryption Paradigm

Let $\mathcal{K}$ be a KEM given by $\mathcal{K}$.keygen, $\mathcal{K}$.encap, and $\mathcal{K}$.decap. Let further $\mathcal{D}$ denote a DEM given by $\mathcal{D}$.encrypt and $\mathcal{D}$.decrypt. We assume that both schemes are compatible in the sense that for all security parameters κ, $\mathcal{K}$'s and $\mathcal{D}$'s key spaces are equal.

$\mathcal{K}$ and $\mathcal{D}$ can be combined as per the *hybrid encryption paradigm* (called also the KEM/DEM paradigm) to build the public-key encryption scheme $(\mathcal{K}, \mathcal{D})$[1] that we describe in Fig. 1.2.

We refer to Herranz et al. (2006) for the necessary and sufficient conditions on the KEM and the DEM to obtain a certain security level for the resulting hybrid encryption scheme.

1.1.4 Tag-Based Encryption (TBE)

Tag-based encryption, also referred to as encryption with labels, was first introduced in Shoup and Gennaro (2002). In these schemes, the encryption algorithm takes as input, in addition to the public key *pk* and the message *m* intended to be encrypted, a tag *t* which specifies information related to the message *m* and to its encryption context. Similarly, the decryption algorithm takes additionally to the ciphertext and the private key the tag under which the ciphertext was created. Security notions are then defined as usual except that the adversary specifies to his challenger the tag to be used in the challenge ciphertext, and in case he (the adversary) is allowed to query oracles, then he cannot query them on the pair consisting of the challenge ciphertext and the tag used to form it. There are also weakened security models for this type of encryption where the adversary specifies the challenge tag before getting the public key of the scheme, and during the game, he (the adversary) is not allowed to query the allowed oracles w.r.t. the challenge tag; we talk in this case about *selective-tag*

[1]Throughout the text, we will use the notation $(\mathcal{K}, \mathcal{D})$ to refer to the public-key encryption scheme resulting from the combination of the KEM $\mathcal{K}$ and the DEM $\mathcal{D}$ using the hybrid encryption paradigm.

security. We specify in the following the formal definition of indistinguishability under a selective-tag and weak chosen-ciphertext attack (IND-st-wCCA).

Definition 1.10 (Selective-Tag Indistinguishability for TBE (IND-st-wCCA)**)** Let Γ be a tag-based encryption scheme. Let further $\mathcal{A}$ denote a PPTM. We consider the following random experiment for security parameter κ and $b \stackrel{R}{\leftarrow} \{0,1\}$:

Experiment $\mathbf{Exp}_{\mathcal{A}}^{\mathsf{IND\text{-}st\text{-}wCCA\text{-}}b}(1^\kappa)$

1. $param \leftarrow \mathtt{setup}(1^\kappa)$
2. $t \leftarrow \mathcal{A}(param)$
3. $(sk, pk) \leftarrow \Gamma.\mathtt{keygen}(param, t, 1^\kappa)$
4. $(m_0, m_1) \leftarrow \mathcal{A}^{\mathtt{decrypt}^{\neg(-,t)}(sk,.)}(param, t, pk)$
5. $e_b \leftarrow \Gamma.\mathtt{encrypt}_{pk}(m_b, t)$
6. $b^\star \leftarrow \mathcal{A}^{\mathtt{decrypt}^{\neg(-,t)}(sk,.)}(e_b)$
7. return $b^\star$

We define the *advantage* of $\mathcal{A}$ *via*:

$$\mathbf{Adv}_{\Gamma,\mathcal{A}}^{\mathsf{IND\text{-}st\text{-}wCCA}}(1^\kappa) = \left| \Pr\left[\mathbf{Exp}_{\Gamma,\mathcal{A}}^{\mathsf{IND\text{-}st\text{-}wCCA\text{-}}b}(1^\kappa) = b\right] - \frac{1}{2} \right|.$$

Given $(t, q_d) \in \mathbb{N}^2$ and $\varepsilon \in [0,1]$, $\mathcal{A}$ is called a (t, ε, q_d)-IND-st-wCCA adversary against Γ if, running in time t and issuing q_d decryption queries, $\mathcal{A}$ has $\mathbf{Adv}_{\Gamma,\mathcal{A}}^{\mathsf{IND\text{-}st\text{-}wCCA}}(1^\kappa) \geq \varepsilon$ in the above experiment. The scheme Γ is said to be (t, ε, q_d)-IND-st-wCCA secure if no (t, ε, q_d)-IND-st-wCCA adversary against it exists.

1.1.4.1 Tag-Based KEMs

A tag-based KEM was introduced in Abe et al. (2005) as a generalization of KEM. Actually, both the encapsulation and decapsulation algorithms in tag-based KEMs input, in addition to the classical arguments, an extra string called a tag.

Security in tag-based KEMs is defined along the same lines of those in tag-based encryption. For instance the adversary has to specify the tag to be used in the challenge, and is not allowed to query the challenge w.r.t. this challenge tag to whatever oracle he has access to. For the sake of completeness, we give the IND-st-wCCA security notion for tag-based KEM.

Definition 1.11 (Selective-Tag Indistinguishability for Tag-Based KEMs) Let $\mathcal{K}$ be a tag-based KEM. Let further $\mathcal{A}$ denote a PPTM. We consider the following random experiment for security parameter κ and $b \stackrel{R}{\leftarrow} \{0,1\}$ (K denotes the key space of $\mathcal{K}$).

Experiment $\mathbf{Exp}_{\mathcal{A}}^{\mathsf{IND\text{-}st\text{-}wCCA\text{-}}b}(1^\kappa)$

1. $param \leftarrow \texttt{setup}(1^\kappa)$
2. $t \leftarrow \mathcal{A}(param)$
3. $(sk, pk) \leftarrow \mathcal{K}.\texttt{keygen}(param, t, 1^\kappa)$
4. $\mathcal{I} \leftarrow \mathcal{A}^{\texttt{decap}^{\neg(-,t)}(sk,.)}(param, t, pk)$
5. $(c^\star, k_1^\star) \leftarrow \mathcal{K}.\texttt{encap}_{pk}(t)$; $k_0^\star \stackrel{R}{\leftarrow} \mathsf{K}$
6. $b^\star \leftarrow \mathcal{A}^{\texttt{decap}^{\neg(-,t)}(sk,.)}(param, pk, c^\star, k_b^\star)$
7. return $b^\star$.

We define the *advantage* of $\mathcal{A}$ *via*:

$$\mathbf{Adv}_{\mathcal{K},\mathcal{A}}^{\mathsf{IND\text{-}st\text{-}wCCA}}(1^\kappa) = \left|\Pr\left[\mathbf{Exp}_{\mathcal{K},\mathcal{A}}^{\mathsf{IND\text{-}st\text{-}wCCA\text{-}}b}(1^\kappa) = b\right] - \frac{1}{2}\right|.$$

Given $(t, q_d) \in \mathbb{N}^2$ and $\varepsilon \in [0, 1]$, $\mathcal{A}$ is called a (t, ε, q_d)-IND-st-wCCA adversary against $\mathcal{K}$ if, running in time t and issuing q_d decapsulation queries, $\mathcal{A}$ has $\mathbf{Adv}_{\Gamma,\mathcal{A}}^{\mathsf{IND\text{-}st\text{-}wCCA}}(1^\kappa) \geq \varepsilon$ in the above experiment. The scheme $\mathcal{K}$ is said to be (t, ε, q_d)-IND-st-wCCA secure if no (t, ε, q_d)-IND-st-wCCA adversary against it exists.

1.1.5 Commitment Schemes

A commitment scheme (Brassard et al. 1988) consists of three algorithms:
$\texttt{keygen}(\mathbf{1}^\kappa)$ This algorithm generates probabilistically a public commitment key pk.
$\texttt{commit}_{pk}(\mathbf{m})$ This is a probabilistic algorithm that, on input a public key pk and a message m, produces a pair (c, r): c serves as the commitment value (locked box), and r as the opening value.
$\texttt{open}_{pk}(\mathbf{m}, \mathbf{c}, \mathbf{r})$ This is a deterministic algorithm that given a pair (c, r) along with a public key pk and an alleged message m, checks whether
$(c, r) \stackrel{?}{=} \texttt{commit}_{pk}(m)$. The algorithm $\texttt{open}$ must succeed if the commitment was correctly formed (correctness).

Typically, we require a commitment scheme to be hiding, binding, and injective; we talk then of a secure commitment.

Definition 1.12 (Hiding in Commitment Schemes) Let Ω be a commitment scheme, and $\mathcal{A}$ denote a PPTM. Consider the following experiment for security parameter κ and $b \stackrel{R}{\leftarrow} \{0, 1\}$

Experiment $\mathbf{Exp}_{\Omega,\mathcal{A}}^{\mathsf{hid}\text{-}b}(\kappa)$

1. $pk \leftarrow \Omega.\texttt{keygen}(\kappa)$
2. $(m_0^\star, m_1^\star, \mathcal{I}) \leftarrow \mathcal{A}(\mathsf{find}, pk)$
3. $(c^\star, r^\star) \leftarrow \Omega.\texttt{commit}_{pk}(m_b^\star)$
4. $d \leftarrow \mathcal{A}(\mathsf{guess}, \mathcal{I}, c^\star)$
5. return d

We define the *advantage* of $\mathcal{A}$ *via*:

$$\mathbf{Adv}^{\mathsf{hid}}_{\Omega,\mathcal{A}}(\kappa) = \left|\Pr\left[\mathbf{Exp}^{\mathsf{hid\text{-}b}}_{\Omega,\mathcal{A}}(\kappa) = b\right] - \frac{1}{2}\right|.$$

Given $t \in \mathbb{N}$ and $\varepsilon \in [0, 1]$, $\mathcal{A}$ is called a (t, ε)-hiding adversary against Ω if, running in time t, $\mathcal{A}$ has $\mathbf{Adv}^{\mathsf{hid}}_{\Omega,\mathcal{A}}(\kappa) \geq \varepsilon$ in the above experiment. The scheme Ω is said to be (t, ε)-hiding if no (t, ε)-hiding adversary against it exists.

Definition 1.13 (Binding in Commitment Schemes) Let Ω be a commitment scheme, and $\mathcal{A}$ denote a PPTM. Consider the following experiment for security parameter κ

Experiment $\mathbf{Exp}^{\mathsf{bind}}_{\Omega,\mathcal{A}}(\kappa)$

1. $pk \leftarrow \Omega.\mathtt{keygen}(\kappa)$
2. $(c, r, r', m, m') \leftarrow \mathcal{A}(pk)$
3. return 1 if and only if :
   - $m \neq m'$
   - $(c, r) = \mathtt{commit}_{pk}(m)$ and $(c, r') = \mathtt{commit}_{pk}(m')$

We define the *advantage* of $\mathcal{A}$ *via*:

$$\mathbf{Adv}^{\mathsf{bind}}_{\Omega,\mathcal{A}}(\kappa) = \Pr\left[\mathbf{Exp}^{\mathsf{bind}}_{\Omega,\mathcal{A}}(\kappa) = 1\right].$$

Given $t \in \mathbb{N}$ and $\varepsilon \in [0, 1]$, $\mathcal{A}$ is called a (t, ε)-binding adversary against Ω if, running in time t, $\mathcal{A}$ has $\mathbf{Adv}^{\mathsf{bind}}_{\Omega,\mathcal{A}}(\kappa) \geq \varepsilon$ in the above experiment. The scheme Ω is said to be (t, ε)-binding if no (t, ε)-binding adversary against it exists.

Definition 1.14 (Injectivity in Commitment Schemes) A commitment scheme Ω is said to be injective if $\Omega.\mathtt{commit}$ for a fixed message (viewed as a function of the opening value) is injective, that is, given a message m, for any two pairs (c, r) and (c', r') produced using $\Omega.\mathtt{commit}$ w.r.t. public key $\Omega.pk$ on m such that $r \neq r'$, we have $c \neq c'$.

We provide in Fig. 1.3 an illustration of an injective, computationally binding and statistically hiding commitment: the Pedersen commitment. Injectivity is straightforward. The commitment is further statistically hiding because r is random in $\mathbb{Z}_d$ and so is $c = g^r y^m$, regardless of m. Besides, the binding property is achieved under the discrete logarithm assumption in $\mathbb{G}$ (see the subsequent section).

Before ending this subsection, we note the similarity between public-key encryption and commitment schemes. Actually, it is not hard to see that indistinguishable (IND-CPA) encryption implies a secure commitment scheme. The main difference is that encryption, contrarily to commitments, requires a trapdoor (private key) that allows to always recover the message from the ciphertext.

setup(1^κ)	: Choose a group ($\mathbb{G} = \langle g \rangle, \cdot$) with prime order d
keygen(1^κ)	: $y \xleftarrow{R} \mathbb{G}$ ($\mathsf{DL}_g(y)$ is unknown) ; $pk \leftarrow y$
commit(m)	: $r \xleftarrow{R} \mathbb{Z}_d$; $c \leftarrow g^r y^m$; return (c,r)
open(c,r,m)	: $c \stackrel{?}{=} g^r y^m$

Fig. 1.3 Pedersen's commitment scheme

1.2 Number-Theoretic Problems

Public-key cryptography is based on the intractability of solving certain computational problems. The two computationally intractable problems upon which is based the security of most public-key cryptosystems are integer factorization and discrete logarithm (DL).

This subsection highlights those two problems and quickly browses through further problems, emanating from them, that are among the frequently used ingredients in modern cryptography.

1.2.1 Factoring-Related Problems

Factoring Factoring consists in splitting an integer into a product of smaller integers. We know that $\mathbb{Z}$ is a unique factorization domain, that is, every natural number factors *uniquely* into a product of prime powers. Therefore, factorization of an integer comes to finding prime powers that divide it (prime divisors along with their multiplicities). Note that once the prime divisors of an integer are found, then it is easy to find the multiplicities by trial divisions. Factoring can consequently be reduced to finding prime divisors of integers. On another note, given an algorithm that computes a non-trivial prime divisor of an integer, the same algorithm can be recursively applied (to the co-factor) to compute the complete factorization of the integer in question.

It is then enough to state the factorization problem as follows: given a product $N = p \times q$ of two equally-sized odd primes such that $p < q$, find p and q. This version is the most common in public-key cryptography, and its presumed difficulty is essentially the same as that of the general version.

The advantage of an adversary $\mathcal{A}$ against the stated factoring problem is defined by:

$$\mathsf{Adv}(\mathcal{A}) = \Pr\left[\begin{array}{r} (p,q,N) \leftarrow \mathtt{keygen}(1^\kappa), \\ p,q \xleftarrow{\tau \text{ operations}} \mathcal{A}(N), \\ N = p \times q. \end{array}\right]$$

where the probability is taken over the random generation of the factorization instance as well as on all the random choices of the adversary $\mathcal{A}$.

Finally, we say that *the factoring assumption holds for security parameter* κ if:

$$\tau = \mathsf{poly}(\kappa) \Rightarrow \mathsf{Adv}(\mathcal{A}) = \mathsf{negl}(\kappa).$$

Root Extraction (RSA) This problem, named after its inventors (Rivest, Shamir, and Adleman), was introduced in Rivest et al. (1978) as a trapdoor one-way function that is suitable for building signature and public-key encryption schemes. It is defined as follows.

Let N be a product of two distinct and equally sized odd primes p and q (p and q are κ-bit integers). Let further y be an integer in $\mathbb{Z}_N^\times$ and $e > 1$ be an integer co-prime with $\phi(N)$. The task of an RSA adversary $\mathcal{A}$ is to compute the unique integer x in $\mathbb{Z}_N^\times$ such that $x^e \equiv y \bmod N$. Clearly the trapdoor is the modular inverse of e (mod $\phi(n)$) which can be efficiently computed using the euclidean algorithm once the factorization of N is known. The rest follows from Euler's theorem.

The advantage of an RSA adversary $\mathcal{A}$ is defined by:

$$\mathsf{Adv}(\mathcal{A}) = \Pr\left[\begin{array}{r} (p,q,N,e) \leftarrow \mathtt{keygen}(1^\kappa), \\ y \xleftarrow{R} \mathbb{Z}_N^\times, \\ x \xleftarrow{\tau \text{ operations}} \mathcal{A}(N,e,y), \\ x^e \equiv y \bmod N. \end{array}\right]$$

where the probability is taken over the random generation of the RSA instance as well as on all the random choices of the RSA adversary.

Finally, we say that *the RSA assumption holds in* $\mathbb{Z}_N^\times$ if:

$$\tau = \mathsf{poly}(\kappa) \Rightarrow \mathsf{Adv}(\mathcal{A}) = \mathsf{negl}(\kappa).$$

1.2.2 Discrete-Log-Related Problems

In this subsection, $(\mathbb{G} = \langle g \rangle, \cdot)$ denotes a finite multiplicative cyclic group generated by g. Unless the contrary is stated, the order d of $\mathbb{G}$ is assumed to be prime.

Discrete Logarithm (DL) The discrete logarithm problem consists in, given an element $y \in \mathbb{G}$, computing x such that $y = g^x$.

It can be easily proven that solving the discrete logarithm problem in a group of order d with known factorization can be efficiently reduced to solving the same problem in groups whose orders are the prime factors of d (see for example Stinson 2006, Chap. 6). This explains why we consider in the literature only groups of prime order.

The advantage of an adversary $\mathcal{A}$ against the DL problem is given by:

$$\mathsf{Adv}(\mathcal{A}) = \Pr\left[\begin{array}{r} (\mathbb{G}, d, g) \leftarrow \texttt{keygen}(1^\kappa), \\ y \xleftarrow{R} \mathbb{G}, \\ x \xleftarrow{\tau \text{ operations}} \mathcal{A}(\mathbb{G}, d, g, y), \\ g^x = y. \end{array}\right]$$

where the probability is taken over the generation of the DL instance as well as on the random choices of $\mathcal{A}$.

Similarly, we say that *the discrete logarithm (DL) assumption holds in* $\mathbb{G}$ if:

$$\tau = \mathsf{poly}(\kappa) \Rightarrow \mathsf{Adv}(\mathcal{A}) = \mathsf{negl}(\kappa).$$

Computational Diffie-Hellman (CDH) The computational Diffie-Hellman problem, first invoked in Diffie and Hellman (1976) consists in, given g^a and g^b for some uniformly chosen a, b from $\mathbb{Z}_d$, computing g^{ab}. The advantage of a CDH adversary $\mathcal{A}$ is given by:

$$\mathsf{Adv}(\mathcal{A}) = \Pr\left[\begin{array}{r} (\mathbb{G}, d, g) \leftarrow \texttt{keygen}(1^\kappa), \\ (a, b) \xleftarrow{R} \mathbb{Z}_d^\times, \\ g^{ab} \xleftarrow{\tau \text{ operations}} \mathcal{A}(\mathbb{G}, d, g, g^a, g^b). \end{array}\right]$$

where the probability is taken over the generation of the CDH instance as well as on the random choices of $\mathcal{A}$.

Similarly, we say that *the Computational Diffie-Hellman (CDH) assumption holds (in* $\mathbb{G}$*)* if:

$$\tau = \mathsf{poly}(\kappa) \Rightarrow \mathsf{Adv}(\mathcal{A}) = \mathsf{negl}(\kappa).$$

It is obvious that solving CDH is easier than computing discrete logarithms. The converse is not obvious, however significant progress has been made towards showing that CDH is almost as hard as the discrete logarithm problem.

Decisional Diffie-Hellman The input to this problem consists of $A = g^a$, $B = g^b$, and $C = g^c$, where a, b are chosen uniformly at random from $\mathbb{Z}_d$ and c is either $ab \bmod d$ or a random element in $\mathbb{Z}_d$. The polynomial-time adversary $\mathcal{A}$ is then requested to decide whether $c \equiv ab \bmod d$ or not. We define the advantage of such an adversary as follows.

$$\mathsf{Adv}(\mathcal{A}) = \left| \Pr\left[\begin{array}{r} (\mathbb{G}, d, g) \leftarrow \texttt{keygen}(1^\kappa), \\ (a, b) \xleftarrow{R} \mathbb{Z}_d^\times, \\ b^\star \xleftarrow{R} \{0, 1\} \\ \text{if } b^\star = 1 \text{ then } c \leftarrow ab \bmod d \text{ else } c \xleftarrow{R} \mathbb{Z}_d^\times, \\ d^\star \xleftarrow{\tau \text{ operations}} \mathcal{A}(\mathbb{G}, d, g, g^a, g^b, g^c), \\ b^\star = d^\star. \end{array} \right] - \frac{1}{2} \right|$$

The probability is taken over the generation of the DDH instance and on the random choices of $\mathcal{A}$.

Similarly, we say that *the Decisional Diffie-Hellman (DDH) assumption holds in* $\mathbb{G}$ if:

$$\tau = \mathsf{poly}(\kappa) \Rightarrow \mathsf{Adv}(\mathcal{A}) = \mathsf{negl}(\kappa).$$

Gap Diffie-Hellman (GDH) The input and output of this problem are similar to those of the CDH problem, with the exception of supporting the adversary $\mathcal{A}$ with a DDH oracle that he can query on any DDH instance of his choosing.

$$\mathsf{Adv}(\mathcal{A}) = \Pr\left[\begin{array}{r} (\mathbb{G}, d, g) \leftarrow \texttt{keygen}(1^\kappa), \\ (a, b) \xleftarrow{R} \mathbb{Z}_d^\times, \\ g^{ab} \xleftarrow{\tau \text{ operations}} \mathcal{A}^{\mathsf{DDH}}(\mathbb{G}, d, g, g^a, g^b). \end{array} \right]$$

where DDH: $(g^a, g^b, g^c) \longmapsto c \stackrel{?}{\equiv} ab \bmod d$, and the probability is taken over the generation of the GDH instance and on the random choices of $\mathcal{A}$.

Similarly, we say that *the Gap Diffie-Hellman (GDH) assumption holds in* $\mathbb{G}$ if:

$$\tau = \mathsf{poly}(\kappa) \Rightarrow \mathsf{Adv}(\mathcal{A}) = \mathsf{negl}(\kappa).$$

The CDH problem is obviously harder than the DDH and GDH problems. There is actually a clear separation between CDH and DDH in the so-called *bilinear groups* (Joux and Nguyen 2003).

Remark 1.3 (Random-Self Reducibility) The CDH, DDH, and GDH problems are *random-self reducible*, i.e. one can generate from a specific instance a random one. Thus, the average case and worst case of all these problems are equivalent.

1.3 Reductionist Security

Reductionist security is a methodology adopted by cryptographers to gain confidence in their systems. Actually, the last two decades have seen the following established mindset for conducting the development of cryptographic systems:

1. Define clearly the security notion the system needs to meet, by combining the security goal the system should attain and the adversarial power the attacker has access to.
2. Describe a well studied problem P upon which the security of the system will rest.
3. Provide a *security reduction* from the studied problem to breaking the scheme in question. That is, provide a polynomial-time algorithm $\mathcal{R}$ that solves the problem P given access to an algorithm $\mathcal{A}$ breaking the security of the system in the sense defined in Step 1. Such a security proof will guarantee the security of the system if the problem P is believed to be hard.

The rest of this section is devoted to unfolding some tools in reductionist security that we will use later in our study.

1.3.1 Cryptographic Reductions

Cryptographic Adversaries An adversary in cryptography is modeled by a probabilistic Turing machine that performs computations in order to break a certain security notion of a given cryptosystem. Adversaries against cryptographic systems are assumed to be computationally bounded, i.e. run in time polynomial in the security parameter.

The success of an adversary at breaking a given cryptosystem (in the sense of a given security notion) is measured by its *advantage* which is specified as a function of the security parameter.

Throughout the text, a function $f(\kappa)$ is said to be *negligible* if $f(\kappa) < 1/p(\kappa)$ for every polynomial $p(\kappa)$ and sufficiently large κ. A function $f(\kappa)$ is called *overwhelming* if $1 - f(\kappa)$ is negligible. Finally, a function is *noticeable* if it is not negligible.

Cryptographic Reductions A reduction in cryptology, often denoted $\mathcal{R}$, is informally an algorithm solving some problem given access to an adversary $\mathcal{A}$ against some cryptosystem. Both the reduction and the adversary are considered probabilistic Turing machines.

To be able to use the adversary $\mathcal{A}$, the reduction $\mathcal{R}$ must simulate $\mathcal{A}$'s environment (instance generation, queries if any...) in a way that is indistinguishable from the real model with noticeable probability, where the probability is taken over

$\mathsf{setup}(1^\kappa)$	: Choose a group $(\mathbb{G} = \langle g \rangle, \cdot)$ with prime order d
$\mathsf{keygen}(1^\kappa)$	: $x \xleftarrow{R} \mathbb{Z}_d$; $y \leftarrow g^x$; $pk \leftarrow (d,g,y)$; $sk \leftarrow (d,g,x)$
$\mathsf{encrypt}(m)$	: $t \xleftarrow{R} \mathbb{Z}_d$; $c_1 \leftarrow g^t$; $c_2 \leftarrow my^t$; return (c_1,c_2)
$\mathsf{decrypt}(c_1,c_2)$	: return $(c_2c_1^{-x})$

Fig. 1.4 ElGamal's encryption scheme

the random generation of the problem instance and over the random tosses of both $\mathcal{A}$ and $\mathcal{R}$. The advantage of the reduction $\mathsf{Adv}(\mathcal{R})$ is by definition the success probability in solving the given instance of the problem; again, the probability is taken over all the random tosses.

A Toy Example: ElGamal's Encryption ElGamal's public-key encryption scheme was invented in 1985; it is outlined in Fig. 1.4. It is not hard to see that this scheme is:

1. OW-CPA secure if the CDH problem is hard, i.e. the CDH assumption holds in $\mathbb{G}$.
2. IND-CPA secure if the DDH problem is hard, i.e. the DDH assumption holds in $\mathbb{G}$.
3. OW-PCA secure if the GDH problem is hard, i.e. the GDH assumption holds in $\mathbb{G}$.

Remark 1.4 It is not difficult to see that the ElGamal encryption, mentioned above, is derived from the hybrid encryption (KEM/DEM) paradigm. Encryption of a message m consists in first generating a key y^k and its encapsulation g^k, then multiplying the generated key with the message m. The ciphertext is formed by the result of this encryption in addition to g^k. Decryption of such a ciphertext first retrieves y^k from g^k using the private key x, then recovers the message m using the key y^k.

Key-Preserving Reductions These reductions refer to a wide and popular class of reductions which supply the adversary with the same public key as its challenge key. In this text, we restrict this notion to a smaller class of reductions.

Definition 1.15 (Key-Preserving Reductions) Let $\mathcal{A}$ be an adversary which solves a problem A that is perfectly reducible to OW-CPA breaking some public-key encryption scheme Γ. Let further $\mathcal{R}$ be a reduction breaking some security notion of Γ w.r.t. a public key pk given access to $\mathcal{A}$. $\mathcal{R}$ is said to be *key-preserving* if it launches $\mathcal{A}$ over her own challenge key pk in addition to some other parameters (chosen freely by her) according to the specification of $\mathcal{A}$.

Such reductions were for instance used in Paillier and Villar (2006) to prove a separation between factoring and IND-CCA breaking some factoring-based encryption schemes in the standard model.

As mentioned in Paillier and Villar (2006), key-preserving reductions are transitive, i.e. if there is a key-preserving reduction from A to B, and another key-preserving reduction from B to C, then there is a key-preserving reduction from A to C. Finally, it is worth noting that reductions among security notions of a given cryptosystem are key-preserving (in the wide sense of the term).

1.3.2 Proof Models

The design of efficient cryptosystems that provide strong security guarantees is unfortunately not easy in practice. In fact, there are very few efficient cryptosystems that rest on difficult problems and whose security proofs do not make any assumptions on the scheme components, i.e. proofs in the so-called "standard model". This has driven two main lines of research in this area.

The first one consists in *idealizing* some components of the cryptosystem subject to the study; the security proof reduces solving the presumed hard problem to breaking the scheme with respect to an adversary accessing the idealized object through an oracle.

One of the most celebrated idealizations is the random oracle model (ROM) which models a random hash function. Recall that a hash function is used to map messages of arbitrary size to *digests* of fixed size; such a property allows to ensure the message integrity and justifies the extensive use of these functions in messages authentication mechanisms. The random oracle methodology treats the hash function as a theoretical black box that responds to every query with a uniformly chosen random string from the output domain, with the exception of giving the same answer to the same query. Random oracles proved useful in cryptography and they were first considered by Fiat and Shamir (1986) to remove interaction from 3-round public-coin identification schemes. Later, they were used by Bellare and Rogaway (1993) to provide generic constructions of encryption and signature schemes. Unfortunately, proofs in the ROM do not provide any insights about the real security when the adversary exploits special properties of the used hash function. This explains the number of results questioning the proven security of certain cryptosystems in the ROM (Goldwasser and Tauman Kalai 2003; Paillier and Vergnaud 2005).

The second approach maintains the standard model but softens the security guarantees of the scheme. The past years have consequently seen the emergence of many unstudied problems along with their strong corresponding assumptions. To gain confidence in these newly introduced problems, researchers often establish their equivalence with well studied problems in very restrictive frameworks, e.g. Schnorr and Jakobsson (2000), Smart (2001), Brown (2005); we are back again to the limitations of proofs in idealized models.

The moral of this tale is straightforward: in cryptographic design, one should try as much as possible to use standard security assumptions, while alleviating or even removing idealized proof models from the security reductions.

1.3.3 Meta-reductions in Cryptography

Meta-reductions are probabilistic oracle (single or multi-oracle) Turing machines, where one oracle tape consists of an efficient reduction from some problem to another. Meta-reductions have been successfully used in a number of important results, e.g. the result in Boneh and Venkatesan (1998) which proves the impossibility of reducing algebraically factoring to RSA, or the results in Paillier and Vergnaud (2005) and Paillier (2007) which show that some well known signatures, which are proven secure in the random oracle, cannot preserve the same security in the standard model. Although most meta-reductions (used in cryptography) apply only to a category of reductions, e.g. key preserving reductions (Paillier 2007; Paillier and Villar 2006) or algebraic reductions (Boneh and Venkatesan 1998; Paillier and Vergnaud 2005), they constitute an efficient tool to separate cryptographic problems (Boneh and Venkatesan 1998) or to disprove that the security of some cryptographic scheme rests on the hardness of some problem.

Figure 1.5 depicts the typical use of a meta-reduction in disproving that a given problem P reduces to breaking a given signature scheme Σ. Actually, let $\mathcal{R}$ be an algorithm that solves an instance of the problem P, using an attacker $\mathcal{A}$ against the signature scheme. Naturally, $\mathcal{R}$ needs to simulate to $\mathcal{A}$ the key generation, the signature, and the verification algorithms of Σ. If one can build an efficient algorithm $\mathcal{M}$ that uses $\mathcal{R}$ to solve an instance of the same problem P (note that such an algorithm needs to simulate to $\mathcal{R}$ the adversary $\mathcal{A}$), then one can conclude on the impossibility of the existence of $\mathcal{R}$. In fact, existence of $\mathcal{M}$ indicates that under the hardness of P, the algorithm $\mathcal{R}$ does not exist; otherwise, if P is easy, then $\mathcal{R}$ might exist, however its work is useless (solving a problem known to be easy).

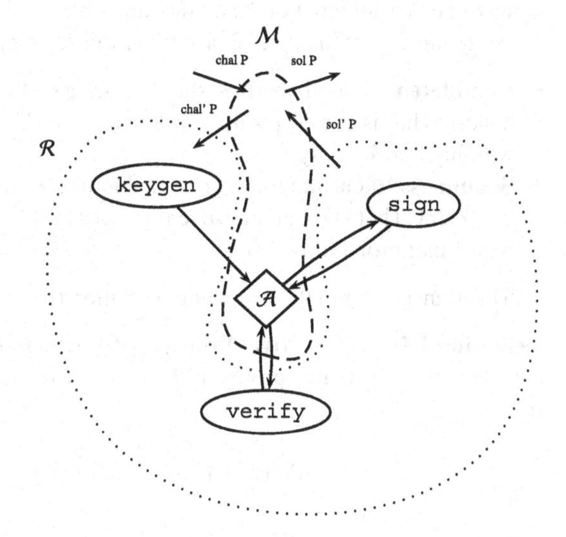

Fig. 1.5 Example of a meta-reduction

1.4 Cryptographic Proof Systems

A basic problem in cryptography consists in conducting a proof, with/out interaction, to convince of the validity of a given statement. The proof must be accepted with high probability when the statement is valid and rejected otherwise. Moreover, it shall not reveal more than the veracity of the proven statement.

In this section, we recall the cryptographic mechanisms that respond to this need, in addition to the different related notions that are relevant for our study.

1.4.1 Interactive Proofs

A model of computation of an interactive proof system was first introduced by Goldwasser et al. (1989). It informally consists of a prover P trying to convince a verifier V that an instance x belongs to a language L. x refers to the common input whereas $(P, V)(x)$ denotes the proof instance carried between P and V at the end of which V is (not) convinced with the membership of x to L:

$$(P, V)(x) \in \{\mathsf{Accept}, \mathsf{Reject}\}$$

P is modeled by a probabilistic Turing machine whereas V is modeled by a *polynomial* probabilistic Turing machine. During $(P, V)(x)$, the parties exchange a sequence of messages called the proof transcript. These messages sizes are polynomial in the size of x. Moreover, $(P, V)(x)$ must terminate in time polynomial in the size of x. The output value $(P, V)(x)$ is a random variable of the common input x, the private input of P and the random coins of both P and V.

We require in an interactive proof the following properties:

- **Completeness:** both parties should successfully run the protocol if they are honest. That is for all positive instances ($x \in L$), we have $(P, V)(x) = \mathsf{Accept}$ with high probability.
- **Soundness:** a cheating prover is unable to convince the verifier with an invalid statement. That is for all negative instances ($x \notin L$), we have $(P, V)(x) = \mathsf{Accept}$ with small probability.

This translates into the following definition (from Mao 2008):

Definition 1.16 (Interactive Proofs (IP)) Let L be a language over a given alphabet. We say that a protocol (P, V) is an interactive proof (IP) system for L if:

$$\Pr\left[(P, V)(x) = \mathsf{Accept} \mid x \in L\right] \geq \epsilon, \tag{1.1}$$

and

$$\Pr\left[(\tilde{P}, V)(x) = \mathsf{Accept} \mid x \notin L\right] \leq \delta \tag{1.2}$$

for every probabilistic Turing machine $\tilde{P}$, where ϵ and δ are constants satisfying

$$\epsilon \in (\frac{1}{2}, 1] \text{ and } \delta \in [0, \frac{1}{2}),$$

where the probability is taken over all the common input values to (P, V) and all random input values of P, $\tilde{P}$, and V.

Soundness Error and Round Efficiency The constant δ in the above definition is called the *soundness error* of the proof. It is always possible to amplify the soundness of a proof until the soundness error becomes a negligibly small quantity (in the security parameter). This can be achieved by repeating the proof many times and accepting only if all proofs verify. After ℓ repetitions, the soundness error δ is reduced to δ^{ℓ}. The protocol is called a *log-round protocol* if the number of repetitions ℓ is linear in the security parameter, and it is called a *poly-round protocol* if ℓ is a polynomial function of the security parameter.

Proofs of Knowledge A proof of knowledge is an interactive proof in which the prover succeeds 'convincing' a verifier that he *knows* something. In addition to the **completeness** property, a proof of knowledge must further satisfy the **validity** or **soundness** property. More precisely, let R be the NP-relation for an NP-language L:

$$L = \{x: \exists \ w \text{ such that } R(x, w) \text{ holds}\}$$

Validity of a proof of knowledge for R captures the intuition that from any (possibly cheating) prover $\tilde{P}$ that is able to convince the verifier with noticeable probability on a statement $x \in L$, there exists an efficient *knowledge extractor* capable of extracting a valid witness for x from $\tilde{P}$ with non-negligible (noticeable) probability. This guarantees that no prover that doesn't know the witness can succeed in convincing the verifier.

1.4.2 Zero-Knowledge (ZK)

In the previous subsection, we exhibited a proof mechanism capable of convincing the verifier with the validity of a valid statement. However, we did not address the question of the additional knowledge the verifier will gain aside from the validity of the statement in question. Ideally, we would like this additional knowledge to be *zero*, thus the name *zero-knowledge*. We define formally this notion as follows (Mao 2008):

Definition 1.17 (Zero-Knowledge) Let (P, V) be an interactive proof system for some language L. We say that (P, V) is **zero-knowledge** if for every $x \in L$, the proof transcript $(P, V)(x)$ can be produced by a probabilistic polynomial-time algorithm (in the size of the input) S with indistinguishable probability distributions:

- if the probability distributions of $(P, V)(x)$ and $S(x)$ are the same, then the protocol (P, V) is said to be perfectly zero-knowledge,
- if the probability distributions of $(P, V)(x)$ and $S(x)$ are statistically indistinguishable, then (P, V) is called a statistical zero-knowledge protocol,
- finally, if the distributions of $(P, V)(x)$ and $S(x)$ are computationally indistinguishable, then (P, V) provides only computational zero-knowledge.

Conventionally, the algorithm S is named a simulator for the ZK protocol since it provides a simulation of the proof transcript. However, in case of perfect ZK protocols, S is called often the equator as it provides a perfect simulation.

Remark 1.5 (Zero-Knowledge Proofs of Knowledge (ZKPoK)) The zero-knowledge property of a proof of knowledge captures the possibility to prove knowledge of the given witness without revealing it. This property is defined, as in interactive proofs, using an efficient simulator, with no access to the prover, capable of producing a proof transcript indistinguishable from the interaction between the genuine prover and the cheating verifier.

A Complexity Theoretic Result: NP ⊂ ZK An important result in complexity theory shows that every language in NP accepts a zero-knowledge proof system. This result has been proven in a constructive manner by first constructing a ZK proof system (P, V) for an NP-complete problem L, e.g. graph 3-colorability by Goldreich et al. (1991) or Boolean satisfiability by Brassard et al. (1988), then propagating this property to the other languages L' in NP as follows:

1. each party computes $x = f(x')$, an instance of the NP-complete language L. It is worth noting that f can by definition be computed and inverted efficiently.
2. P conducts a ZK proof with V to prove that $x \in L$.

It is obvious that the above construction of a ZK proof system for any language in NP constitutes only a theoretic result. In fact, a practical ZK protocol should have the number of interactions (round number) between P and V bounded by a linear function in the security parameter. This cannot be achieved by the above construction since we do not know any linear transformation (reduction) of an NP language to an NP-complete one.

Sequential vs Concurrent Zero-Knowledge We addressed in the preceding subsection the possibility of repeating many times a proof in order to reduce its soundness error. This repetition can be sequential or parallel. The natural question to ask is whether the zero-knowledge feature is preserved or not. The good news is that zero-knowledge is closed under sequential repetition of the protocol (see Goldreich 2001, Chap. 4, Paragraph 4.3.4 for the proof), which means that we can indefinitely reduce the soundness error of a protocol without compromising its zero-knowledgeness. Parallel composition is not however guaranteed to preserve zero-knowledge. Less is the concurrent composition which generalizes both sequential and parallel composition; in this composition, many instances of the protocol are invoked at arbitrary times and proceed at arbitrary pace. This composition turns out to be of significant importance in many real-life applications. Fortunately, there exists a result (Damgård 2000) which shows that a wide range of known zero-knowledge protocols, e.g. Σ protocols (subsequent subsection), can be modified with negligible loss of efficiency to preserve zero-knowledgeness under concurrent composition. A more general (and theoretic) result (Dwork et al. 2004) demonstrates that every NP language accepts a concurrent ZK proof system.

The Auxiliary String Model The auxiliary string model was introduced in Damgård (2000) and it captures the assumption that an auxiliary string with prescribed distribution is available to both the prover and the verifier.

More specifically, an interactive proof in this model consists, in addition to the prover P and the verifier V, a setup algorithm that inputs the security parameter and outputs an auxiliary string σ. Soundness means that no prover can cheat the verifier with noticeable probability where the probability is taken over the choice of the auxiliary string as well as the random tosses of both the prover and the verifier. Zero-knowledge in this setting means that the verifier's entire view, namely the interaction with the prover in addition to the auxiliary string, can be efficiently simulated.

It is worth noting that the auxiliary string model is equivalent to the *common reference string model* used in the study of non-interactive zero-knowledge proofs (Sect. 1.4.4).

Finally, the auxiliary string model can be efficiently implemented in practice in a public-key setting. In fact, public-key infrastructures (PKI) can successfully emulate the algorithm that generates the auxiliary string. We refer to Damgård (2000) for the details.

1.4.3 Σ Protocols

A *public-coin protocol* is an interactive proof in which the verifier chooses all its messages randomly from publicly known sets. A *three-move protocol* can be written in a canonical form in which the messages exchanged in the three moves are often called commitment, challenge, and response respectively. The protocol is said to have the *honest-verifier zero-knowledge property (HVZK)* if there exists an algorithm that is able, provided the verifier behaves as prescribed by the protocol, to produce, without the knowledge of the secret, transcripts that are indistinguishable from those of the real protocol. The protocol is said to have the *special soundness (SpS) property* if there exists an algorithm that is able to extract the secret from two accepting transcripts of the protocol with the same commitment and different challenges. Finally, a three-move public-coin protocol with HVZK and SpS properties is called a *Σ protocol.*

Throughout the document, $\mathtt{Cmt}$, $\mathtt{Rsp}$, and $\mathtt{Dcd}$ denote respectively the algorithms used to compute the commitment, the response, and the decision in a Σ protocol (see Fig. 1.6). The transcript tr of the Σ protocol depicted in Fig. 1.6

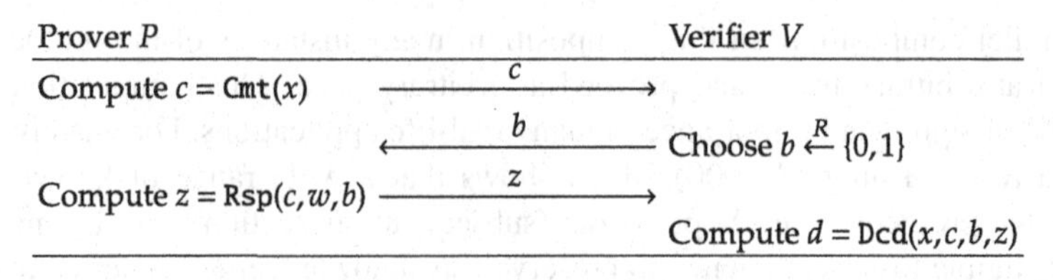

Fig. 1.6 Proof of membership to the language $\{x: (x, w) \in \mathcal{R}\}$ Common input: x and Private input : w

is formed from the messages exchanged between the prover and the verifier, namely (c, b, z). Finally, we will denote by `trSim` the algorithm that simulates the transcript of a Σ protocol on a given instance.

1.4.4 Non-interactive Proofs

Non-interactive proof systems were introduced in Blum et al. (1988). They consist of three entities: a prover, a verifier, and a uniformly selected *common reference string*—crs—(which can be thought of as being selected by a trusted authority TA). Both the verifier and the prover can read the reference string. The interaction consists of a single message sent from the prover to the verifier, who is left with the final decision. The zero-knowledge requirement refers to a simulator that outputs pairs that should be indistinguishable from the pairs (crs, prover's message). It is worth noting here that the definition of zero-knowledge for these proofs is simplified because the verifier cannot affect the prover's actions.

Removing Interaction in Proofs The most famous technique to obtain NIZK (non-interactive zero-knowledge) proofs from their interactive variants is known as the Fiat-Shamir paradigm (Fiat and Shamir 1986). It consists of letting the prover compute the verifier's challenge himself as a hash of the statement to be proved and of the first message. The security of this construction is provided only in the random oracle model (ROM), which constitutes its major shortcoming. In fact, it is not in general possible to instantiate the random oracle with a concrete function and have the security properties preserved. To alleviate the shortcomings of the Fiat-Shamir transform, Lindell (2014) recently proposed a similar transform that turns any Σ protocol for a relation R into a NIZK proof for the associated language L_R in the common reference string model; the new transform achieves two advantages over the original one: (1) the zero-knowledge property holds in the *standard model*, (2) the soundness property holds in the *non-programmable random oracle*, which is an intuitive, less constrained, and thus conceptually preferable model than the original (fully programmable) random oracle.

Another technique, due to Damgård et al. (2006), transforms a three-move interactive ZK protocol with linear answer to a non-interactive ZK one (NIZK) in a registered key model, i.e. in a model where the verifier registers his key. The key ingredient in this transform is a homomorphic encryption scheme Γ that operates on integer values in a suitable range:

$$\Gamma.\texttt{encrypt}_{\Gamma.pk}(m + m') = \Gamma.\texttt{encrypt}_{\Gamma.pk}(m) \cdot \Gamma.\texttt{encrypt}_{\Gamma.pk}(m'),$$

for all messages m and m'. More precisely, let a be the first message computed by the prover, $c \in \mathbb{N}$ be the challenge sent by the verifier, and finally let $z = u + cv$ be the answer computed by the prover in the third step, where $u, v \in \mathbb{N}$. If the verifier chooses a key pair $(\Gamma.pk, \Gamma.sk)$ and publishes an encryption e of the challenge c, then the prover can compute a as usual, $\Gamma.\texttt{encrypt}(z)$ as $\Gamma.\texttt{encrypt}(u)e^v$, and send these quantities to the verifier in one pass. The verifier decrypts $\Gamma.\texttt{encrypt}(z)$ to obtain z and checks whether (a, c, z) is an accepting transcript. The authors in Damgård et al. (2006) proposed an efficient illustration using Paillier's encryption and the proof of equality of two discrete logarithms.

Groth-Sahai's Proof System The Groth-Sahai (GS) (2008) proof system gives efficient non-interactive zero-knowledge (NIZK) proofs for a number of languages that cover many algebraic statements. The principle of the GS proof system consists in using the setup parameters and the common reference string (crs) to commit to the witness components, then generate proof elements that these values committed to satisfy the equations underlying the statement to be proved. The GS proof system could be instantiated under different (mild) assumptions, e.g. the DDH assumption.

References

Abe M, Gennaro R, Kurosawa K, Shoup V (2005) Tag-KEM/DEM: a new framework for hybrid encryption and a new analysis of Kurosawa-Desmedt KEM. In: Cramer R (ed) EUROCRYPT. LNCS, vol 3494. Springer, Heidelberg, pp 128–146

Bellare M, Rogaway P (1993) Random Oracles are practical: a paradigm for designing efficient protocols. In: Denning D, Pyle R, Ganesan R, Sandhu R, Ashby V (eds) Proceedings of the first ACM conference on computer and communications security. ACM Press, New York, pp 62–73

Bellare M, Desai A, Pointcheval D, Rogaway P (1998) Relations among notions of security for public-key encryption schemes. In: Krawczyk H (ed) Advances in cryptology - CRYPTO'98. LNCS, vol 1462. Springer, Heidelberg, pp 26–45

Blum M, Feldman P, Micali S (1988) Non-interactive zero-knowledge and its applications (extended abstract). In: Simon J (ed) STOC. ACM Press, New York, pp 103–112

Boneh D, Venkatesan R (1998) Breaking RSA may not be equivalent to factoring. In: Nyberg K (ed) Advances in cryptology - EUROCRYPT'98. LNCS, vol 1403. Springer, Heidelberg, pp 59–71

Brassard G, Chaum D, Crépeau C (1988) Minimum disclosure proofs of knowledge. J Comput Syst Sci 37(2):156–189

Brown DRL (2005) Generic groups, collision resistance, and ECDSA. Des Codes Cryptogr 35(1):119–152

Damgård I (2000) Efficient concurrent zero-knowledge in the auxiliary string model. In: Preneel B (ed) EUROCRYPT. LNCS, vol 1807. Springer, Heidelberg, pp 418–430
Damgård I, Fazio N, Nicolosi A (2006) Non-interactive zero-knowledge from homomorphic encryption. In: Halevi S, Rabin T (eds) TCC 2006. LNCS, vol 3876. Springer, Heidelberg, pp 41–59
Diffie W, Hellman ME (1976) New directions in cryptography. IEEE Trans Inf Theory 22:644–654
Dolev D, Dwork C, Naor M (1991) Non-malleable cryptography (extended abstract). In: STOC. ACM Press, New York, pp 542–552
Dwork C, Naor M, Sahai A (2004) Concurrent zero-knowledge. J Assoc Comput Mach 51(6): 851–898
Fiat A, Shamir A (1986) How to prove yourself: practical solutions to identification and signature problems. In: Odlyzko AM (ed) CRYPTO. LNCS, vol 263. Springer, Heidelberg, pp 186–194
Goldreich O (2001) Foundations of cryptography. Basic tools. Cambridge University Press, Cambridge
Goldreich O, Micali S, Wigderson A (1991) Proofs that yield nothing but their validity or all languages in NP have zero-knowledge proof systems. J Assoc Comput Mach 38(3):691–729
Goldwasser S, Micali S (1984) Probabilistic encryption. J Comput Syst Sci 28:270–299
Goldwasser S, Tauman Kalai Y (2003) On the (in)security of the Fiat-Shamir Paradigm. In: Sudan M (ed) Proceedings of the 44th IEEE symposium on foundations of computer science (FOCS 2003). IEEE Computer Society, Cambridge, pp 102–113
Goldwasser S, Micali S, Rivest RL (1988) A digital signature scheme secure against adaptive chosen-message attacks. SIAM J Comput 17(2):281–308
Goldwasser S, Micali S, Rackoff C (1989) The knowledge complexity of interactive proof-systems. SIAM J Comput 18(1):186–206
Groth J, Sahai A (2008) Efficient non-interactive proof systems for bilinear groups. In: Smart NP (ed) EUROCRYPT 2008. LNCS, vol 4965. Springer, Heidelberg, pp 415–432
Herranz J, Hofheinz D, Kiltz E (2006) KEM/DEM: necessary and sufficient conditions for secure hybrid encryption. Available at http://eprint.iacr.org/2006/265.pdf
Joux A, Nguyen K (2003) Separating decision Diffie-Hellman from computational Diffie-Hellman in cryptographic groups. J Cryptol 16(4):239–247
Lindell Y (2014) An efficient transform from sigma protocols to NIZK with a CRS and non-programmable random Oracle. IACR Cryptology ePrint Archive 2014:710
Mao W (2008) Modern cryptography: theory & practice. Dorling Kindersley, Noida
Paillier P (2007) Impossibility proofs for RSA signatures in the standard model. In: Abe M (ed) CT-RSA. LNCS, vol 4377. Springer, Heidelberg, pp 31–48
Paillier P, Vergnaud D (2005) Discrete-log based signatures may not be equivalent to discrete-log. In: Roy B (ed) Advances in cryptology - ASIACRYPT 2005. LNCS, vol 3788. Springer, Heidelberg, pp 1–20
Paillier P, Villar J (2006) Trading one-wayness against chosen-ciphertext security in factoring-based encryption. In: Lai X, Chen K (eds) ASIACRYPT. LNCS, vol 4284. Springer, Heidelberg, pp 252–266
Rivest RL, Shamir A, Adleman LM (1978) A method for obtaining digital signatures and public-key cryptosystems. Commun ACM 21:120–126
Schnorr CP, Jakobsson M (2000) Security of signed ElGamal encryption. In: Okamoto T (ed) Advances in cryptology - ASIACRYPT 2000. LNCS, vol 1976. Springer, Heidelberg, pp 73–89
Shoup V, Gennaro R (2002) Securing threshold cryptosystems against chosen ciphertext attack. J Cryptol 15(2):75–96. Earlier version in EUROCRYPT 1998
Smart NP (2001) The exact security of ECIES in the generic group model. In: Honary B (ed) Cryptography and coding, 8th IMA international conference. LNCS, vol 2260. Springer, Heidelberg, pp 73–84
Stinson D (2006) Cryptography: theory and practice. Chapman & Hall/CRC, Taylor and Francis, Boca Raton

Chapter 2
Case-Study Primitives

Abstract This chapter introduces the primitives subject to the study, namely designated-confirmer signatures and signcryption. The presentation covers the syntax of the mentioned primitives in addition to their security properties. Since establishing a formal security model for a cryptographic system is a real challenge and divergence between cryptographers, we subject the model we adhere to to an in-depth comparison with the already established ones; our goal is to have well-reasoned and stringent security properties which capture various attack scenarios.

2.1 Convertible Designated-Confirmer Signatures (CDCS)

2.1.1 Motivation

Digital signatures capture most properties met by signatures in the paper world, for instance the universal verification. However, in some applications, this property is not desired or at least needs to be controlled. The typical applications where we wish to restrain the holder of a signature from convincing other parties of the validity of the signature are:

Licensing software (Chaum and van Antwerpen 1990) A software vendor is willing to embed signatures in his products such that only the paying customers are entitled to check the authenticity of these products. Moreover, he does not wish these paying customers to convince other parties of the genuineness of his goods.

Contract signing (Goldwasser and Waisbard 2004) An employer issues a job offer to a certain candidate. Naturally, the employer needs to compete with the other job offers in order to attract the good candidate. Therefore, he does not wish the offer to be revealed to his competitors. At the same time, the candidate needs more than a verbal or unsigned agreement in order to protect himself from the employer not keeping his promise. Finally, when the candidate accepts the offer, the employer wishes to *convert* the job offer he has issued to a publicly verifiable one, instead of having to issue a new contact.

Undeniable signatures were introduced in Chaum and van Antwerpen (1990) to respond to the aforementioned requirements; they proved critical in situations where privacy or anonymity is a big concern, e.g. licensing software (Chaum and

L. El Aimani, *Verifiable Composition of Signature and Encryption*,
https://doi.org/10.1007/978-3-319-68112-2_2

van Antwerpen 1990), electronic cash (Boyd and Foo 1998; Chaum and Pedersen 1993; Pointcheval 2001) and electronic voting and auctions. In these signatures, the verification can be only attained by means of a cooperation with the signer, called the confirmation/denial protocols. Unfortunately, this very virtue (verification with only the signer's help) became their major shortcoming for many practical applications. The flaw was later repaired in Chaum (1995) by introducing the concept of *designated-confirmer signatures*. In fact, this concept involves three entities, namely the signer who produces the signature, the designated confirmer who confirms or denies the alleged signature, and finally the recipient of the signature. Actually, in the literature, there is a clear separation between designated-confirmer signatures or confirmer signatures for brevity, and *directed signatures* (Lim and Lee 1993) which share the same concept as confirmer signatures with the exception of allowing both the signer and the confirmer to confirm/deny signatures. Finally, a desirable property in confirmer signatures is the convertibility of the signatures to ordinary ones. Indeed, such a property turned out to play a central role in fair payment protocols (Boyd and Foo 1998).

2.1.2 Syntax

A convertible designated-confirmer signature (CDCS) scheme consists of the following procedures:

$\mathtt{setup}(1^\kappa)$ On input a security parameter κ, generate probabilistically the public parameters *param* of the scheme. Although not always explicitly mentioned, *param* serves as an input to all the algorithms/protocols that follow.

$\mathtt{keygen}_{\mathsf{E}}(1^\kappa, param)$ This probabilistic algorithm outputs the key pair $(pk_{\mathsf{E}}, sk_{\mathsf{E}})$ for entity E in the system; E can either be the signer S who issues the confirmer signatures, or the confirmer C who confirms/denies the signatures.

$\mathtt{sign}_{sk_{\mathsf{S}}}(m, pk_{\mathsf{C}})$ On input sk_{S}, pk_{C} and a message m, this probabilistic algorithm outputs a confirmer signature μ on m.

$\mathtt{verify}_{\{coins \vee sk_{\mathsf{C}}\}}(\mu, m, pk_{\mathsf{S}}, pk_{\mathsf{C}})$ This is an algorithm, run by the signer on a *just generated* signature or by the confirmer on *any* signature. The input to the algorithm is the alleged signature μ, the message m, pk_{S}, pk_{C} in addition to the coins *coins* used to produce the signature if the algorithm is run by the signer, or sk_{C} if it is run by the confirmer. The output is either 1 if the signature if valid, or 0 otherwise.

$\mathtt{sconfirm}_{\langle \mathsf{S}(coins_\mu), \mathsf{V}\rangle}(\mu, m, pk_{\mathsf{S}}, pk_{\mathsf{C}})$ This is an interactive protocol where the signer S convinces a verifier V of the validity of a signature he has just generated. The common input comprises the signature and the message in question, in addition to pk_{S} and pk_{C}. The private input of S consists of the random coins used to produce the signature μ on m.

$\mathtt{confirm/deny}_{\langle \mathsf{C}(sk_{\mathsf{C}}), \mathsf{V}\rangle}(\mu, m, pk_{\mathsf{S}}, pk_{\mathsf{C}})$ These are interactive protocols between the confirmer C and a verifier V. Their common input consists of

pk_S, pk_C, the alleged signature μ, and the message m. The confirmer uses sk_C to convince the verifier of the validity/invalidity of the signature μ on m. At the end, the verifier accepts or rejects the proof.

$\texttt{convert}_{sk_C}(\mu, m, pk_S, pk_C)$ This is an algorithm run by the confirmer C using sk_C, in addition to pk_C and pk_S, on a potential confirmer signature μ and some message m. The result is either $\perp$ if μ is not a valid confirmer signature on m, or a string σ which is a valid digital signature on m w.r.t. pk_S.

$\texttt{verifyconverted}\,(\sigma, m, pk_S)$ This is an algorithm for verifying converted signatures. It inputs the converted signature σ, the message m and pk_S and outputs either 0 or 1.

Remark 2.1 (Notation) For the sake of simplicity, the public/private keys as well as the private coins will be often omitted from the description of the above algorithms/protocols. Therefore, whenever the context is clear, `sign`, `sconfirm`, {`confirm`, `deny`}, `convert`, and `verifyconverted` will only involve the message and the corresponding confirmer/converted signature.

Remark 2.2 In Gentry et al. (2005) and Wang et al. (2007), the authors give the possibility of obtaining *directly* digital signatures on any given message. We find this unnecessary since it is already enough that a CDCS scheme supports the convertibility feature.

2.1.3 Security Model for CDCS

Since their introduction, many definitions and security models for CDCS have emerged. Recall that a security property is, as commonly agreed on, an attribute allowing a cryptographic scheme to withstand malicious attempts aiming at make it deviate from its prescribed task. These malicious attempts can be classified into two categories:

1. Attempts conducted by adversaries *inside* the system. This is for instance the case where the scheme operators are dishonest, coerced, or where they simply have their private keys compromised or stolen.
2. Attempts conducted by adversaries *outside* the system. These are the default attacks any cryptographic scheme should take into consideration.

A cryptographic scheme resilient against the first type of attacks is said to procure security in the *insider model*, whereas a scheme resilient against the second type of attacks is said to be secure in the *outsider model*. Consideration of the appropriate security model depends upon the functionality of the scheme; for some schemes it is enough to consider outsider security, for others it is imperative to consider insider security at least for some scheme properties.

The rest of this section will be devoted to the definition of the security properties we opt for in this text, as well as to the comparison of these properties with the popular ones found in the literature.

2.1.3.1 Completeness

Every signature produced by `sign` should be validated by the algorithm `verify` and correctly converted. Moreover, valid signatures should be correctly confirmed by `sconfirm` and `confirm`, and invalid signatures should be correctly denied by `deny` if the signer S and the confirmer C follow honestly the protocols.

Definition 2.1 (Completeness in CDCS) Let *CS* be a CDCS scheme with signer S and confirmer C. Let further (pk_S, sk_S) and (pk_C, sk_C) denote the signer and the confirmer key pairs resp., and m be a message from the message space of *CS*. We consider the following experiment where V is a PPTM.

Experiment $\mathbf{Exp}_{CS}^{\text{completeness}}(m, pk_S, pk_C)$

1. $\mu \leftarrow CS.\mathtt{sign}_{sk_S}(m, pk_C)$
2. $\psi \xleftarrow{R} CS.\text{space}: CS.\mathtt{verify}_{sk_C}(\psi, m, pk_C, pk_S) = 0$
3. $out_0 \leftarrow CS.\mathtt{verify}_{\{coins_\mu \vee sk_C\}}(\mu, m, pk_C, pk_S)$
4. $\sigma \leftarrow CS.\mathtt{convert}_{sk_C}(\mu, m)$
5. $out_1 \leftarrow CS.\mathtt{verifyconverted}_{pk_S}(\sigma, m)$
6. $\langle done \mid out_2 \rangle \leftarrow CS.\mathtt{sconfirm}_{\langle \mathsf{S}(coins_\mu), \mathsf{V} \rangle}(\mu, m, pk_S, pk_C)$
7. $\langle done \mid out_3 \rangle \leftarrow CS.\mathtt{confirm}_{\langle \mathsf{C}(sk_C), \mathsf{V} \rangle}(\mu, m, pk_S, pk_C)$
8. $\langle done \mid out_4 \rangle \leftarrow CS.\mathtt{deny}_{\langle \mathsf{C}(sk_C), \mathsf{V} \rangle}(\psi, m, pk_S, pk_C)$
9. return $(out_0 \wedge out_1, out_2 \wedge out_3 \wedge out_4)$

The scheme *CS* complete if, for all signer's and confirmer's key pairs (pk_S, sk_S) and (pk_C, sk_C) resp., for all messages m, the first outcome of Experiment $\mathbf{Exp}_{CS}^{\text{completeness}}(m, pk_S, pk_C)$ is always 1, and the second outcome is 1 with overwhelming probability, where the probability is taken over all the random choices.

2.1.3.2 Soundness

This property informally means that an adversary who compromises the private keys of both the signer and the confirmer cannot convince the verifier of the validity (invalidity) of an invalid (a valid) confirmer signature.

Definition 2.2 (Soundness in CDCS) Let *CS* be a CDCS scheme with signer S and confirmer C. Let further $\mathcal{A}$ and V be PPTMs. Consider the following experiment for security parameter κ.

Experiment $\mathbf{Exp}_{CS,\mathcal{A}}^{\text{soundness}}(1^\kappa)$

1. $param \leftarrow \mathtt{setup}(1^\kappa)$
2. $(pk_S, sk_S) \leftarrow CS.\mathtt{keygen_S}(1^\kappa)$; $(pk_C, sk_C) \leftarrow CS.\mathtt{keygen_C}(1^\kappa)$
3. $(m, \psi) \leftarrow \mathcal{A}(sk_S, sk_C, coins_\psi): CS.\mathtt{verify}_{\{coins_\psi \vee sk_C\}}(\psi, m, pk_C, pk_S) = 0$
4. $(m, \mu) \leftarrow \mathcal{A}(sk_S, sk_C, coins_\mu): CS.\mathtt{verify}_{\{coins_\mu \vee sk_C\}}(\mu, m, pk_S, pk_C) = 1$
5. $\langle done \mid out_1 \rangle \leftarrow CS.\mathtt{sconfirm}_{\langle \mathcal{A}(sk_S, sk_C, coins_\psi), \mathsf{V} \rangle}(\psi, m, pk_S, pk_C)$
6. $\langle done \mid out_2 \rangle \leftarrow CS.\mathtt{confirm}_{\langle \mathcal{A}(sk_S, sk_C, coins_\psi), \mathsf{V} \rangle}(\psi, m, pk_S, pk_C)$

7. $\langle done \mid out_3 \rangle \leftarrow \mathtt{deny}_{\langle \mathcal{A}(sk_S, sk_C, coins_\mu), \mathsf{V} \rangle}(\mu, m, pk_S, pk_C)$
8. return $(out_1 \vee out_2 \vee out_3)$

CS is sound if the success probability of any polynomial-time adversary $\mathcal{A}$ (returning 1) in Experiment $\mathbf{Exp}^{\mathsf{soundness}}_{CS,\mathcal{A}}(1^\kappa)$ is negligible in κ; the probability is taken over all the random tosses.

2.1.3.3 Non-transferability

This property captures the simulatability of `sconfirm`, `confirm`, and `deny`. It is defined through the following games which involve the adversary, the signer and the confirmer of the CDCS scheme CS, and a simulator:

Game 1 the adversary $\mathcal{A}$ is given the public keys of the signer and of the confirmer, namely pk_S and pk_C resp. He can then make arbitrary queries of type $\{\mathtt{sign}, \mathtt{sconfirm}\}$ to the signer and of type $\{\mathtt{confirm}, \mathtt{deny}\}$ and `convert` to the confirmer. Eventually, the adversary presents two strings m and μ for which he wishes to carry out, on the common input (m, μ, pk_S, pk_C), the protocol `sconfirm` with the signer (if μ has been just generated by the signer on m), or the protocols $\{\mathtt{confirm}, \mathtt{deny}\}$ with the confirmer. The private input of the signer is the randomness used to generate the signature μ (in case μ is a signature just generated by the signer), whereas the private input of the confirmer is his private key sk_C. The adversary continues issuing queries to both the signer and the confirmer until he decides that this phase is over and produces an output.

Game 2 this game is similar to the previous one with the difference of playing a simulator instead of running the real signer or the real confirmer when it comes to the interaction of the adversary with the signer in `sconfirm` or with the confirmer in $\{\mathtt{confirm}, \mathtt{deny}\}$ on the common input (μ, m, pk_S, pk_C). The simulator is not given the private input of neither the signer nor the confirmer. It is however allowed to issue a single oracle call that tells whether μ is a valid confirmer signature on m w.r.t. pk_S and pk_C. Note that the simulator in this game refers to a probabilistic polynomial-time Turing machine with rewind.

The confirmer signatures are said to be non-transferable if there exists an efficient simulator such that for all (pk_S, pk_C), the outputs of the adversary in **Game 1** and **Game 2** are indistinguishable. In other words, the adversary should not be able to tell whether he is playing **Game 1** or **Game 2**.

Definition 2.3 (Non-transferability in CDCS) Let CS be a CDCS scheme with signer S and confirmer C, and let $\mathcal{A}$ be a PPTM. Consider the following experiment where κ is a security parameter.

Experiment $\mathbf{Exp}^{\mathsf{non-transferability}}_{\{CS,\mathcal{A}\}}(1^\kappa)$

1. $param \leftarrow CS.\mathtt{setup}(1^\kappa)$
2. $(pk_S, sk_S) \leftarrow CS.\mathtt{keygen_S}(1^\kappa)$; $(pk_C, sk_C) \leftarrow CS.\mathtt{keygen_C}(1^\kappa)$

3. $(m, \mu) \leftarrow \mathcal{A}^{\mathfrak{S},\mathfrak{Cv},\mathfrak{V}}(pk_S, pk_C)$
   $\mathfrak{S} : m_i \longmapsto CS.\mathtt{sign}_{sk_S}(m_i, pk_C)$
   $\mathfrak{Cv} : (\mu_i, m_i) \longmapsto CS.\mathtt{convert}_{sk_C}(\mu_i, m_i, pk_S, pk_C)$
   $\mathfrak{V} : (\mu_i, m_i) \longmapsto CS.\{\mathtt{sconfirm}, \mathtt{confirm}, \mathtt{deny}\}(\mu_i, m_i, pk_S, pk_C)$
4. $b \xleftarrow{R} \{0,1\}$
   if $b = 1$ then $\langle done \mid out_0 \rangle \leftarrow \mathtt{prove}_{\langle \mathsf{P(w)}, \mathcal{A} \rangle}(\mu, m, pk_S, pk_C)$;
   if $b = 0$ then $\langle done \mid out_1 \rangle \leftarrow \mathtt{prove}_{\langle \mathsf{Sim}, \mathcal{A} \rangle}(\mu, m, pk_S, pk_C)$;
   if prove $= CS.\mathtt{sconfirm}$ then $\{\mathsf{P} = \mathsf{S}\,;\, w = coins_\mu\}$
   if prove $\in CS.\{\mathtt{confirm}, \mathtt{deny}\}$ then $\{\mathsf{P} = \mathsf{C}\,;\, w = sk_C\}$
5. $b^\star \leftarrow \mathcal{A}^{\mathfrak{S},\mathfrak{Cv},\mathfrak{V}}(\mu, m, pk_S, pk_C)$
6. return $(b = b^\star)$

We define the advantage of the adversary $\mathcal{A}$ as:

$$\mathbf{Adv}^{\text{non-transferability}}_{\{CS,\mathcal{A}\}}(\kappa) = \left| \Pr\left[\mathbf{Exp}^{\text{non-transferability}}_{\{CS,\mathcal{A}\}}(1^\kappa) = 1 \right] - \frac{1}{2} \right|$$

The confirmer signatures are said to be non-transferable if there exists an efficient simulator Sim such that for all (pk_S, pk_C), the advantage of any polynomial-time adversary $\mathcal{A}$ in the above experiment is negligible in κ. The probability is taken over all the random coins.

Online vs Offline Non-transferability The definition of non-transferability is the same adopted in Camenisch and Michels (2000), Gentry et al. (2005), and Wang et al. (2007). In particular, it thrives on the *concurrent* zero knowledgeness of the {sconfirm, confirm, deny} protocols, and guarantees only the so-called *offline non-transferability*.

In fact, non-transferability is not preserved, as remarked by Liskov and Micali (2008), if the verifier interacts concurrently with the prover and with an unexpected verifier.

One way to circumvent this shortcoming consists in requiring the mentioned protocols to be *designated-verifier proofs (Jakobsson et al. 1996)*, i.e. require the verifier to be able to efficiently provide the proofs underlying sconfirm, confirm, and deny, such that no efficient adversary is able to tell whether he is interacting with the genuine prover or with the verifier. This approach was adhered to for instance in Chow and Haralambiev (2011) and Monnerat and Vaudenay (2011). Similarly, many of the realizations of confirmer signatures, proposed later in the text, satisfy also this stronger notion of *online non-transferability* since the protocols {sconfirm, confirm, deny} can be easily turned into Σ protocols which can be in turn transformed efficiently into designated-verifier proofs.

2.1.3.4 Unforgeability

It is defined through a game between a challenger and an adversary $\mathcal{A}$; the adversary gets the signer's public key of a CDCS scheme, and generates the confirmer key pair. $\mathcal{A}$ is further allowed to query the signer on polynomially many messages, say q_s, for confirmer signatures. At the end, $\mathcal{A}$ outputs a pair consisting of a message $m^\star$, that has not been queried before, and a string $\mu^\star$. $\mathcal{A}$ wins the game if $\mu^\star$ is a valid confirmer signature on $m^\star$.

Definition 2.4 (Unforgeability in CDCS (EUF-CMA)**)** Let CS be a CDCS scheme, and let $\mathcal{A}$ be a PPTM. Consider the following experiment where κ is a security parameter.

Experiment $\mathbf{Exp}_{CS,\mathcal{A}}^{\mathsf{EUF\text{-}CMA}}(1^\kappa)$

1. $param \leftarrow \mathtt{setup}(1^\kappa)$
2. $(pk_\mathsf{S}, sk_\mathsf{S}) \leftarrow CS.\mathtt{keygen}_\mathsf{S}(1^\kappa)$
3. $(pk_\mathsf{C}, sk_\mathsf{C}) \leftarrow \mathcal{A}(pk_\mathsf{S})$
4. $(m^\star, \mu^\star) \leftarrow \mathcal{A}^{\mathfrak{S}}(pk_\mathsf{S}, pk_\mathsf{C}, sk_\mathsf{C})$
   $\mathfrak{S} : m \longmapsto CS.\mathtt{sign}_{sk_\mathsf{S}}(m, pk_\mathsf{C})$
5. return 1 if and only if:
   - $\mathtt{verify}(\mu^\star, m^\star, pk_\mathsf{S}, pk_\mathsf{C}) = 1$
   - $m^\star$ was not queried to $\mathfrak{S}$

We define the *advantage* of $\mathcal{A}$ *via*:

$$\mathbf{Adv}_{CS,\mathcal{A}}^{\mathsf{EUF\text{-}CMA}}(\kappa) = \Pr\left[\mathbf{Exp}_{CS,\mathcal{A}}^{\mathsf{EUF\text{-}CMA}}(\kappa) = 1\right].$$

We say that a CDCS scheme CS is (t, ϵ, q_s)-EUF-CMA secure if there is no adversary, operating in time t and issuing q_s signature queries, that wins the game in the above experiment with advantage greater than ϵ, where the probability is taken over all the random choices.

*Remark 2.3 (Strong Unforgeability (*SEUF-CMA*))* Similarly to digital signatures, a SEUF-CMA adversary against confirmer signatures is given access to a confirmer signature oracle that can be adaptively queried on any message, and is asked to produce a *new* confirmer signature on some message that is *allowed to be previously queried*. A confirmer signature is SEUF-CMA secure if no SEUF-CMA attacker, operating in polynomial-time, against it exists.

Remark 2.4 (Insider Security (for the Signer) Against Malicious Confirmers*)* The above unforgeability definition considers, similarly to Camenisch and Michels (2000), Gentry et al. (2005), Wang et al. (2007), and Wikström (2007) the *insider security model against malicious confirmers*, i.e. the adversary is *allowed* to choose his key pair $(sk_\mathsf{C}, pk_\mathsf{C})$. This is justified by the need of preventing the confirmer from impersonating the signer by issuing valid signatures on his behalf.

2.1.3.5 Invisibility

Invisibility against a chosen-message attack (INV-CMA) is defined through the following game between an attacker $\mathcal{A}$ and his challenger C: after $\mathcal{A}$ gets the public parameters of the CDCS scheme CS from C, he starts **Phase 1** where he queries the `sign`, `sconfirm`, `confirm`, `deny`, and `convert` oracles in an adaptive way. Once $\mathcal{A}$ decides that **Phase 1** is over, he outputs two messages $m_0^\star, m_1^\star$ as challenge messages. C picks uniformly at random a bit $b \stackrel{R}{\leftarrow} \{0,1\}$. Then $\mu^\star$ is generated using the signing oracle on the message $m_b^\star$. Next, $\mathcal{A}$ starts adaptively querying the previous oracles (**Phase 2**), with the exception of not querying $(\mu^\star, m_i^\star), i = 0, 1$, to the `sconfirm`, {`confirm`, `deny`}, and `convert` oracles. At the end, $\mathcal{A}$ outputs a bit $b^\star$. He wins the game if $b = b^\star$.

Definition 2.5 (Invisibility in CDCS (INV-CMA)**)** Let CS be a CDCS scheme and $\mathcal{A}$ be a PPTM. Consider the following experiment for security parameter κ.

Experiment $\mathbf{Exp}_{CS,\mathcal{A}}^{\mathsf{INV\text{-}CMA}}(1^\kappa)$

1. $param \leftarrow \mathtt{setup}(1^\kappa)$
2. $(pk_S, sk_S) \leftarrow CS.\mathtt{keygen_S}(1^\kappa)$; $(pk_C, sk_C) \leftarrow CS.\mathtt{keygen_C}(1^\kappa)$
3. $(m_0^\star, m_1^\star, \mathcal{I}) \leftarrow \mathcal{A}^{\mathfrak{S},\mathfrak{Cv},\mathfrak{V}}(param, pk_S, pk_C)$
   - $\mathfrak{S} : m \longmapsto CS.\mathtt{sign}_{sk_S}(m, pk_C)$
   - $\mathfrak{Cv} : (\mu, m) \longmapsto CS.\mathtt{convert}_{sk_C}(\mu, m, pk_S, pk_C)$
   - $\mathfrak{V} : (\mu, m) \longmapsto CS.\{\mathtt{sconfirm}, \mathtt{confirm}, \mathtt{deny}\}_{\{coins_\mu \vee sk_C\}}(\mu, m)$
4. $b \stackrel{R}{\leftarrow} \{0,1\}$; $\mu^\star \leftarrow CS.\mathtt{sign}_{sk_S}(m_b^\star, pk_C)$
5. $b^\star \leftarrow \mathcal{A}^{\mathfrak{S},\mathfrak{Cv},\mathfrak{V}}(\text{guess}, \mathcal{I}, \mu^\star, pk_S, pk_C)$
   - $\mathfrak{S} : m \longmapsto CS.\mathtt{sign}_{sk_S}(m, pk_C)$
   - $\mathfrak{Cv} : (\mu, m)(\neq (\mu^\star, m_i^\star), i = 0, 1) \longmapsto CS.\mathtt{convert}_{sk_C}(\mu, m)$
   - $\mathfrak{V} : (\mu, m)(\neq (\mu^\star, m_i^\star), i = 0, 1) \longmapsto CS.\{\mathtt{sconfirm}, \mathtt{confirm}, \mathtt{deny}\}(\mu, m)$
6. return $(b = b^\star)$

We define $\mathcal{A}$'s advantage as

$$\mathsf{Adv}_{CS,\mathcal{A}}^{\mathsf{INV\text{-}CMA}}(\kappa) = \left|\Pr[\mathbf{Exp}_{CS,\mathcal{A}}^{\mathsf{INV\text{-}CMA}}(1^\kappa) = 1] - \frac{1}{2}\right|.$$

A CDCS scheme is $(t, \epsilon, q_s, q_v, q_{sc})$-INV-CMA secure if no adversary operating in time t, issuing q_s queries to the signing oracle (followed potentially by queries to the `sconfirm` oracle), q_v queries to the confirmation/denial oracles and q_{sc} queries to the selective conversion oracle that wins the above game with advantage greater that ϵ. The probability is taken over all the coin tosses.

Insider vs Outsider Invisibility The invisibility notion is considered in the *outsider security model*, that is, the adversary does not know the private key of the signer.

Actually, insider privacy is needed in very limited applications; the typical example, given in An et al. (2002), is when the adversary happens to steal the private key of the signer, but we still wish to protect the privacy of the recorded confirmer signatures signed by the genuine signer. However, even in this very situation, there seems to be no tangible extra power that an insider attacker can gain from having access to the signing key.

In fact, the insider adversary in the definitions in Gentry et al. (2005) and Wang et al. (2007) is not allowed to ask the verification/conversion of valid confirmer signatures on the challenge messages (otherwise his task would be trivial). This restriction involves for instance confirmer signatures that the adversary may have forged on the challenge messages using the stolen signing key. Let's see how this can translate into a real attack scenario: the insider adversary $\mathcal{A}$ has compromised the signer's key and wishes to break the invisibility of an alleged signature μ on some message m. Naturally, the restriction imposed earlier can be achieved by the signer revoking his key and alerting the confirmer not to verify/convert confirmer signatures involving the message m and potentially further messages. The revocation of the signing key implies also not considering signatures that have been issued after the key has been compromised. This leaves $\mathcal{A}$ with only verification/conversion queries on messages where the corresponding signatures have been issued by the genuine signer before the revocation of the signing key. This seems to reduce $\mathcal{A}$'s adversarial power down to that of an outsider attacker.

Bottom line is outsider invisibility might be all that one needs and *can have* in practice. We provide anyway in Chap. 9 tools to achieve insider privacy at the expense of a small extra-cost in terms of bandwidth and computation.

Invisibility vs Non-transferability The outlined invisibility notion does not guarantee non-transferability of the resulting signatures. In fact, the confirmer signature might convince the recipient that the signer was involved in the signature of some message. We refer to the discussion in Gentry et al. (2005, Sect. 3) for techniques that can be used by the signer to camouflage the presence of valid signatures. Later, we will propose constructions that achieve a stronger notion of invisibility; for such confirmer signatures, it is difficult to distinguish a valid confirmer signature on some message from a random string sampled from the signature space.

2.2 Signcryption

2.2.1 Motivation and Challenges

Signcryption is a primitive, introduced by Zheng (1997), to simultaneously perform the functions of both signature and encryption in a way that is more efficient than signing and encrypting separately. In fact, many real-life applications entail both the confidentiality and the authenticity/integrity of the transmitted data; an illustrative example is electronic elections where the voter wants to encrypt his vote to guarantee privacy, and at the same time, the voting center needs to ensure that the encrypted vote comes from the entity that claims its authorship.

Since the introduction of this primitive, many constructions which achieve different levels of security have been proposed. On a high level, security of a signcryption scheme involves two properties; privacy and authenticity. Privacy is analogous to indistinguishability in encryption schemes, and it denotes the infeasibility to infer any information about the *signcrypted* message. Authenticity is similar to unforgeability in signature schemes and it denotes the difficulty to impersonate the *signcrypter*. Defining formally those two properties is a fundamental divergence in signcryption constructions as there are many issues which come into play:

- **Two-user versus multi-user setting** In the two-user setting, adopted for instance in An et al. (2002), a single sender (the entity that creates the signcryption) interacts with a single receiver (the entity that recovers the message from the signcryption). Although such a setting is too simplistic to represent the reality, e.g. the case of electronic elections, it provides however an important preliminary step towards modeling and building schemes in the multi-user setting. We refer to Chap. 8 for tweaks that allow to derive multi-user security from two-user security.
- **Insider versus outsider security** Another consequential difference between security models is whether the adversary is external or internal to the system. The former case corresponds to outsider security, e.g. Dent (2005), whereas the latter denotes insider security which protects the system protagonists even when some of their fellows are malicious or have compromised/lost their private keys (An et al. 2002; Matsuda et al. 2009). It is naturally possible to mix these notions into one single signcryption scheme, i.e. insider indistinguishability and outsider unforgeability (An et al. 2002; Chiba et al. 2011), or outsider indistinguishability and insider unforgeability (Baek et al. 2007). However, the most frequent mix is the latter as illustrated by the number of works in the literature, e.g. An et al. (2002), Jeong et al. (2002), Lee and Lim (2003), and Baek et al. (2007); it is also justified by the necessity to protect the sender from anyone trying to impersonate him including entities in the system. Insider indistinguishability is by contrast needed in very limited applications as described earlier for the confirmer signature case. We refer again to Chap. 9 for techniques that upgrade indistinguishability so as to protect against insider adversaries.
- **Verifiability** A further requirement on signcryption is verifiability which consists in the possibility to prove efficiently the validity of a given signcryption, or to prove that a signcryption has indeed been produced on a given message. In fact, if we consider the example of electronic elections, the voting center might require from the voter a proof of validity of the "signcrypted" vote. Also, the trusted party (the receiver) that decrypts the vote might be compelled, for instance to resolve some later disputes, to prove that the sender has indeed produced the vote; therefore, it would be desirable to support the prover with efficient means to provide such a proof without having to disclose his private input. This property is also needed in filtering out spams in a secure email system. Although a number of constructions (Bao and Deng 1998; Chow et al. 2003; Lee and Lim 2003; Ma 2006; Selvi et al. 2010; Shin et al. 2002) have tackled the notion of verifiability (this notion is often referred to in the literature as public verifiability, and it denotes the possibility to release (by the receiver) some information which allows to publicly verify a signcryption with/out revealing the message in question), most of these schemes do not allow the sender to prove the validity of the created signcryption, nor allow the receiver to prove

without revealing any information, ensuring consequently non-transferability, to a third party, the validity of a signcryption w.r.t. a given message. It is worth noting that the former need, namely, allowing the sender to prove the validity of a signcryption without revealing the message, already manifests in the IACR electronic voting scheme (The Helios voting scheme) where the sender proves the validity of the encrypted vote to the voting manager. The scheme nonetheless does not respond to the formal security requirements of a signcryption scheme.

2.2.2 Syntax

A verifiable signcryption scheme consists of the following algorithms/protocols:

`setup`(1^κ) This probabilistic algorithm inputs a security parameter κ, and generates the public parameters *param* of the signcryption scheme.

`keygen`$_U(1^\kappa, param)$, $\mathsf{U} \in \{\mathsf{S}, \mathsf{R}\}$ This probabilistic algorithm inputs the security parameter κ and the public parameters *param*, and outputs a key pair $(pk_\mathsf{U}, sk_\mathsf{U})$ for the system user U which is either the sender S or the receiver R.

`signcrypt`$(m, sk_\mathsf{S}, pk_\mathsf{S}, pk_\mathsf{R})$ This probabilistic algorithm inputs a message m, the key pair $(sk_\mathsf{S}, pk_\mathsf{S})$ of the sender, the public key pk_R of the receiver, and outputs the signcryption μ of the message m.

`proveValidity`$(\mu, pk_\mathsf{S}, pk_\mathsf{R})$ This is an interactive protocol between the receiver or the sender who has just generated a signcryption μ on some message, and any verifier: the sender uses the randomness used to create μ (as private input) and the receiver uses his private key sk_R in order to convince the verifier that μ is a valid signcryption on some message. The common input to both the prover and the verifier comprises the signcryption μ, pk_S, and pk_R. At the end, the verifier either accepts or rejects the proof.

`unsigncrypt`$(\mu, sk_\mathsf{R}, pk_\mathsf{R}, pk_\mathsf{S})$ This is a deterministic algorithm which inputs a putative signcryption μ on some message, the key pair $(sk_\mathsf{R}, pk_\mathsf{R})$ of the receiver, and the public key pk_S of the sender, and outputs either the message underlying μ or an error symbol $\perp$.

$\{$`confirm`, `deny`$\}(\mu, m, pk_\mathsf{R}, pk_\mathsf{S})$ These are interactive protocols between the receiver and any verifier; the receiver uses his private key sk_R (as private input) to convince any verifier that a signcryption μ on some message m is/is not valid. The common input comprises the signcryption μ and the message m, in addition to pk_R and pk_S. At the end, the verifier is either convinced of the validity/invalidity of μ w.r.t. m or not.

`sigExtract`$(\mu, m, sk_\mathsf{R}, pk_\mathsf{R}, pk_\mathsf{S})$ This is an algorithm which inputs a signcryption μ, a message m, the key pair $(sk_\mathsf{R}, pk_\mathsf{R})$ of the receiver, and the public key pk_S of the sender, and outputs either an error symbol $\perp$ if μ is not a valid signcryption on m, or a string σ which is a valid digital signature on m w.r.t pk_S otherwise.

`sigVerify`$(\sigma, m, pk_{\mathsf{S}})$ This is an algorithm for verifying extracted signatures. It inputs the extracted signature σ, the message m and pk_{S}, and outputs either 0 or 1.

2.2.3 Security Model

We require in a signcryption scheme correctness (completeness) and soundness. Moreover, the protocols `proveValidity` and {`confirm`, `deny`} must be complete, sound, and non-transferable. The formal definitions of these notions are similar to the confirmer signatures case and thus will omitted.

Finally, we require in a signcryption scheme two further properties: unforgeability, which protects the sender's authenticity from *malicious insider* adversaries (i.e. the receiver), and indistinguishability, which protects the sender's privacy from *outsider adversaries.*

Definition 2.6 (Unforgeability for Signcryption (EUF-CMA)) We consider a signcryption scheme SC given by the algorithms/protocols defined earlier in this section. Let $\mathcal{A}$ be a PPTM. We consider the following random experiment with security parameter κ:

Experiment $\mathbf{Exp}_{SC,\mathcal{A}}^{\mathsf{EUF\text{-}CMA}}(1^\kappa)$

1. $param \leftarrow SC.\mathtt{setup}(1^\kappa)$
2. $(pk_{\mathsf{S}}, sk_{\mathsf{S}}) \leftarrow SC.\mathtt{keygen}_S(1^\kappa, param)$
3. $(pk_{\mathsf{R}}, sk_{\mathsf{R}}) \leftarrow \mathcal{A}(pk_{\mathsf{S}})$
4. $(m^\star, \mu^\star) \leftarrow \mathcal{A}^{\mathfrak{S}}(pk_{\mathsf{S}}, pk_{\mathsf{R}}, sk_{\mathsf{R}})$
   $\mathfrak{S} : m \longmapsto SC.\mathtt{signcrypt}\{sk_{\mathsf{S}}, pk_{\mathsf{S}}, pk_{\mathsf{R}}\}(m)$
5. return 1 if and only if :
   - $SC.\mathtt{unsigncrypt}_{\{sk_{\mathsf{R}},pk_{\mathsf{R}},pk_{\mathsf{S}}\}}[\mu^\star] = m^\star$
   - $m^\star$ was not queried to $\mathfrak{S}$

We define the *advantage* of $\mathcal{A}$ *via*:

$$\mathsf{Adv}_{SC,\mathcal{A}}^{\mathsf{EUF\text{-}CMA}}(1^\kappa) = \Pr\left[\mathbf{Exp}_{SC,\mathcal{A}}^{\mathsf{EUF\text{-}CMA}}(1^\kappa) = 1\right],$$

where the probability is taken over all the coin tosses.

Given $(t, q_s) \in \mathbb{N}^2$ and $\varepsilon \in [0, 1]$, $\mathcal{A}$ is called a (t, ε, q_s)-EUF-CMA adversary against SC if, running in time t and issuing q_s queries to the $SC.$`signcrypt` oracle, $\mathcal{A}$ has advantage greater than ε. The scheme SC is said to be (t, ε, q_s)-EUF-CMA secure if no (t, ε, q_s)-EUF-CMA adversary against it exists.

Remark 2.5 Note that $\mathcal{A}$ is not given the `proveValidity`, `unsigncrypt`, `sigExtract`, and {`confirm`, `deny`} oracles. In fact, these oracles are useless for him as he has the receiver's private key sk_{R} at his disposal.

*Remark 2.6 (Strong Unforgeability (*SEUF-CMA*))* Again, a SEUF-CMA adversary against a signcryption scheme is allowed to forge on queried messages. A signcryption scheme is SEUF-CMA secure if no SEUF-CMA attacker against it exists.

Definition 2.7 (Indistinguishability for Signcryption (IND-CCA)) Let SC be a signcryption scheme, and let $\mathcal{A}$ be a PPTM. We consider the random experiment for $b \xleftarrow{R} \{0,1\}$ and security parameter κ.

Experiment $\mathbf{Exp}_{SC,\mathcal{A}}^{\mathsf{IND\text{-}CCA\text{-}}b}(1^\kappa)$

1. $param \leftarrow SC.\mathtt{setup}(1^\kappa)$
2. $(sk_\mathsf{S}, pk_\mathsf{S}) \leftarrow SC.\mathtt{keygen}_S(1^\kappa, param)$
3. $(sk_\mathsf{R}, pk_\mathsf{R}) \leftarrow SC.\mathtt{keygen}_\mathsf{R}(1^\kappa, param)$
4. $(m_0^\star, m_1^\star, \mathcal{I}) \leftarrow \mathcal{A}^{\mathfrak{S},\mathfrak{V},\mathfrak{U},\mathfrak{C}}(pk_\mathsf{S}, pk_\mathsf{R})$;
   - $\mathfrak{S} : m \longmapsto SC.\mathtt{signcrypt}_{\{sk_\mathsf{S}, pk_\mathsf{S}, pk_\mathsf{R}\}}(m)$
   - $\mathfrak{V} : \mu \longmapsto SC.\mathtt{proveValidity}(\mu, pk_\mathsf{S}, pk_\mathsf{R})$
   - $\mathfrak{U} : \mu \longmapsto SC.\mathtt{unsigncrypt}_{\{sk_\mathsf{R}, pk_\mathsf{R}, pk_\mathsf{S}\}}(\mu)$
   - $\mathfrak{C} : (\mu, m) \longmapsto SC.\{\mathtt{confirm}, \mathtt{deny}\}(\mu, m, pk_\mathsf{R}, pk_\mathsf{S})$
   - $\mathfrak{P} : (\mu, m) \longmapsto SC.\mathtt{sigExtract}(\mu, m, pk_\mathsf{R}, pk_\mathsf{S})$
5. $\mu^\star \leftarrow SC.\mathtt{signcrypt}_{\{sk_\mathsf{S}, pk_\mathsf{S}, pk_\mathsf{R}\}}(m_b^\star)$
6. $d \leftarrow \mathcal{A}^{\mathfrak{S},\mathfrak{V},\mathfrak{U},\mathfrak{C}}(\mathcal{I}, \mu^\star, pk_\mathsf{S}, pk_\mathsf{C})$
   - $\mathfrak{S} : m \longmapsto SC.\mathtt{signcrypt}_{\{sk_\mathsf{S}, pk_\mathsf{S}, pk_\mathsf{R}\}}(m)$
   - $\mathfrak{V} : \mu \longmapsto SC.\mathtt{proveValidity}(\mu, pk_\mathsf{S}, pk_\mathsf{R})$
   - $\mathfrak{U} : \mu (\neq \mu^\star) \longmapsto SC.\mathtt{unsigncrypt}_{\{sk_\mathsf{R}, pk_\mathsf{R}, pk_\mathsf{S}\}}(\mu)$
   - $\mathfrak{C} : (\mu, m)(\neq (\mu^\star, m_i^\star), i = 0, 1) \longmapsto SC.\{\mathtt{confirm}, \mathtt{deny}\}(\mu, m)$
   - $\mathfrak{P} : (\mu, m)(\neq (\mu^\star, m_i^\star), i = 0, 1) \longmapsto SC.\mathtt{sigExtract}(\mu, m)$
7. return (d)

We define the *advantage* of $\mathcal{A}$ *via:*

$$\mathbf{Adv}_{SC,\mathcal{A}}^{\mathsf{IND\text{-}CCA}}(1^\kappa) = \left| \Pr\left[\mathbf{Exp}_{SC,\mathcal{A}}^{\mathsf{IND\text{-}CCA\text{-}}b}(1^\kappa) = b \right] - \frac{1}{2} \right|,$$

where the probability is taken over all the random choices.

Given $(t, q_s, q_v, q_u, q_{cd}, q_e) \in \mathbb{N}^6$ and $\varepsilon \in [0,1]$, $\mathcal{A}$ is called a $(t, \varepsilon, q_s, q_v, q_u, q_{cd}, q_e)$-IND-CCA adversary against SC if, running in time t and issuing q_s queries to the `signcrypt` oracle, q_v queries to the `proveValidity` oracle, q_u queries to the `unsigncrypt` oracle, q_{cd} queries to the {`confirm`, `deny`} oracle, and q_e to the `sigExtract` oracle, $\mathcal{A}$ has advantage greater than ε. The scheme SC is $(t, \varepsilon, q_s, q_v, q_u, q_{cd}, q_e)$-IND-CCA secure if no $(t, \varepsilon, q_s, q_v, q_u, q_{cd}, q_e)$-IND-CCA adversary against it exists.

References

An JH, Dodis Y, Rabin T (2002) On the security of joint signature and encryption. In: Knudsen LR (ed) Advances in cryptology - EUROCRYPT 2002. LNCS, vol 2332. Springer, Heidelberg, pp 83–107

Baek J, Steinfeld R, Zheng Y (2007) Formal proofs for the security of signcryption. J Cryptol 20(2):203–235

Bao F, Deng RH (1998) A signcryption scheme with signature directly verifiable by public key. In: Imai H, Zheng Y (eds) Public key cryptography. LNCS, vol 1431. Springer, Heidelberg, pp 55–59

Boyd C, Foo E (1998) Off-line fair payment protocols using convertible signatures. In: Ohta K, Pei D (eds) Advances in cryptology - ASIACRYPT'98. LNCS, vol 1514. Springer, Heidelberg, pp 271–285

Camenisch J, Michels M (2000) Confirmer signature schemes secure against adaptative adversaries. In: Preneel B (ed) Advances in cryptology - EUROCRYPT 2000. LNCS, vol 1807. Springer, Heidelberg, pp 243–258

Chaum D (1995) Designated confirmer signatures. In: De Santis A (ed) Advances in cryptology - EUROCRYPT'94. LNCS, vol 950. Springer, Heidelberg, pp 86–91

Chaum D, Pedersen TP (1993) Wallet databases with observers. In: Brickell EF (ed) Advances in cryptology - CRYPTO'92. LNCS, vol 740. Springer, Heidelberg, pp 89–105

Chaum D, van Antwerpen H (1990) Undeniable signatures. In: Brassard G (ed) Advances in cryptology - CRYPTO'89. LNCS, vol 435. Springer, Heidelberg, pp 212–216

Chiba D, Matsuda T, Schuldt JN, Matsuura K (2011) Efficient generic constructions of signcryption with insider security in the multi-user setting. In: Lopez J, Tsudik G (eds) Applied cryptography and network security. LNCS, vol 6715. Springer, Heidelberg, pp 220–237

Chow SSM, Haralambiev K (2011) Non-interactive confirmer signatures. In: Kiayias A (ed) CT-RSA. LNCS, vol 6558. Springer, Heidelberg, pp 49–64

Chow SM, Yiu SM, Hui L, Chow KP (2003) Efficient forward and provably secure ID-based signcryption scheme with public verifiability and public ciphertext authenticity. In: Lim JI, Lee DH (eds) ICISC. LNCS, vol 2971. Springer, Heidelberg, pp 352–369

Dent AW (2005) Hybrid signcryption schemes with outsider security. In: Zhou J, Lopez J, Deng RH, Bao F (eds) ISC. LNCS, vol 3650. Springer, Heidelberg, pp 203–217

Gentry C, Molnar D, Ramzan Z (2005) Efficient designated confirmer signatures without random oracles or general zero-knowledge proofs. In: Roy B (ed) Advances in cryptology - ASIACRYPT 2005. LNCS, vol 3788. Springer, Heidelberg, pp 662–681

Goldwasser S, Waisbard E (2004) Transformation of digital signature schemes into designated confirmer signature schemes. In: Naor M (ed) Theory of cryptography, TCC 2004. LNCS, vol 2951. Springer, Heidelberg, pp 77–100

Jakobsson M, Sako K, Impagliazzo R (1996) Designated verifier proofs and their applications. In: Maurer UM (ed) Advances in cryptology - EUROCRYPT'96. LNCS, vol 1070. Springer, Heidelberg, pp 143–154

Jeong I, Jeong H, Rhee H, Lee D, Lim J (2002) Provably secure encrypt-then-sign composition in hybrid signcryption. In: Lee PJ, Lim CH (eds) (2003) Information security and cryptology - ICISC 2002, 5th international conference, Seoul, 28–29 November 2002. LNCS, vol 2587. Springer, Heidelberg, pp 16–34

Lee PJ, Lim CH (eds) (2003) Information security and cryptology - ICISC 2002, 5th international conference, Seoul, 28–29 November 2002. LNCS, vol 2587. Springer, Heidelberg. Revised Papers

Lim CH, Lee PJ (1993) Modified Maurer-Yacobi's scheme and its applications. In: Seberry J, Zheng Y (eds) Advances in cryptology - AUSCRYPT '92. LNCS, vol 718. Springer, Heidelberg, pp 308–323

Liskov M, Micali S (2008) Online-untransferable signatures. In: Cramer R (ed) Public key cryptography. LNCS, vol 4939. Springer, Heidelberg, pp 248–267

Ma C (2006) Efficient short signcryption scheme with public verifiability. In: Lipmaa H, Yung M, Lin D (eds) Inscrypt. LNCS, vol 4318. Springer, Heidelberg, pp 118–129

Matsuda T, Matsuura K, Schuldt J (2009) Efficient constructions of signcryption schemes and signcryption composability. In: Roy B, Sendrier N (eds) IndoCrypt, vol 5922. Springer, Berlin/Heidelberg, pp 321–342

Monnerat J, Vaudenay S (2011) Short undeniable signatures based on group homomorphisms. J Cryptol 24(3):545–587

Pointcheval D (2001) Self-scrambling anonymizers. In: Frankel Y (ed) Financial cryptography, 4th international conference, FC 2000. LNCS, vol 1962. Springer, Heidelberg, pp 259–275

Selvi S, Vivek S, Pandu Rangan P (2010) Identity based public verifiable signcryption scheme. In: Heng SH, Kurosawa K (eds) ProvSec. LNCS, vol 6402. Springer, Heidelberg, pp 244–260

Shin JB, Lee K, Shim K (2002) New DSA-verifiable signcryption schemes. In: Lee PJ, Lim CH (eds) (2003) Information security and cryptology - ICISC 2002, 5th international conference, Seoul, 28–29 November 2002. LNCS, vol 2587. Springer, Heidelberg, pp 35–47

Wang G, Baek J, Wong DS, Bao F (2007) On the generic and efficient constructions of secure designated confirmer signatures. In: Okamoto T, Wang X (eds) PKC 2007. LNCS, vol 4450. Springer, Heidelberg, pp 43–60

Wikström D (2007) Designated confirmer signatures revisited. In: Vadhan SP (ed) TCC 2007. LNCS, vol 4392. Springer, Heidelberg, pp 342–361

Zheng Y (1997) Digital signcryption or how to achieve cost(signature & encryption) $\ll$ cost(signature) + cost(encryption). In: Kaliski Jr BS (ed) Advances in cryptology - CRYPTO'97. LNCS, vol 1294. Springer, Heidelberg, pp 165–179

Part II
The “Sign_then_Encrypt” (StE) Paradigm

Chapter 3
Analysis of StE

Abstract StE consists, in case of confirmer signatures, in first signing the message, then encrypting the resulting signature. In case of signcryption, the encryption is conducted on both the message and the produced signature. The construction was first formally (The idea without proof was already known, for instance, it was mentioned in Damgård and Pedersen (New convertible undeniable signature schemes. In: Maurer UM (ed) Advances in cryptology - EUROCRYPT'96. LNCS, vol 1070. Springer, Heidelberg, pp 372–386, 1996).) described for confirmer signatures in Camenisch and Michels (Confirmer signature schemes secure against adaptative adversaries. In: Preneel B (ed) Advances in cryptology - EUROCRYPT 2000. LNCS, vol 1807. Springer, Heidelberg, pp 243–258, 2000), and it suffered the resort to concurrent zero-knowledge (ZK) proofs of general NP statements in the confirmation/denial protocol (i.e. proving knowledge of the decryption of a ciphertext, and that this decryption forms a valid signature on the given message). In this chapter, we analyze the exact security of StE; i.e. we specify the necessary and sufficient assumptions on the components that lead to secure constructions. We examine, on the way, the conjectured security of a celebrated confirmer signature derived from StE, which was left as open problem for more than a decade. Although the results are all stated for confirmer signatures, they can be straightforwardly extended to the signcryption case.

3.1 StE for Confirmer Signatures

3.1.1 The StE Paradigm

Consider the following primitives:

1. **A digital signature scheme** Σ given by: (1) $\Sigma.\mathtt{keygen}$ which generates a key pair ($\Sigma.sk$, $\Sigma.pk$) (2) $\Sigma.\mathtt{sign}$ (3) $\Sigma.\mathtt{verify}$.
2. **A public-key encryption scheme** Γ described by: (1) $\Gamma.\mathtt{keygen}$ that generates the key pair ($\Gamma.sk$, $\Gamma.pk$) (2) $\Gamma.\mathtt{encrypt}$ (3) $\Gamma.\mathtt{decrypt}$.

Recall that the notation $\Gamma.\mathtt{encrypt}_{\{\Gamma.pk,coins\}}(m)$ refers to the ciphertext obtained from encrypting the message m under the public key $\Gamma.pk$ using the random coins *coins* (`encrypt` is a probabilistic algorithm).

L. El Aimani, *Verifiable Composition of Signature and Encryption*,
https://doi.org/10.1007/978-3-319-68112-2_3

$$
\begin{array}{ll}
\mathit{CS}.\mathtt{setup}(1^\kappa) & : \Sigma.\mathtt{setup}(1^\kappa)\,;\ \Gamma.\mathtt{setup}(1^\kappa) \\
\mathit{CS}.\mathtt{keygen}_S(1^\kappa) & : \Sigma.\mathtt{keygen}(1^\kappa) \\
\mathit{CS}.\mathtt{keygen}_C(1^\kappa) & : \Gamma.\mathtt{keygen}(1^\kappa) \\
\mathit{CS}.\mathtt{sign}(m) & : \Gamma.\mathtt{encrypt}_{\{\Gamma.pk,coins\}}(\Sigma.\mathtt{sign}_{\Sigma.sk}(m)) \\
[\mathit{CS}.\mathtt{sconfirm}(\mu,m) & : \mathsf{ZKP}\Big\{(\sigma,coins):\ \mu = \Gamma.\mathtt{encrypt}_{\{\Gamma.pk,coins\}}(\sigma) \quad \wedge \\
 & \qquad \Sigma.\mathtt{verify}_{\Sigma.pk}(\sigma,m) = 1 \Big\} \\
\mathit{CS}.\mathtt{confirm}(\mu,m) & : \mathsf{ZKP}\Big\{(\sigma,\Gamma.sk):\ \sigma = \Gamma.\mathtt{decrypt}_{\Gamma.sk}(\mu) \wedge \Sigma.\mathtt{verify}_{\Sigma.pk}(\sigma,m) = 1\Big\} \\
\mathit{CS}.\mathtt{deny}(\mu,m) & : \mathsf{ZKP}\Big\{(\sigma,\Gamma.sk):\ \sigma = \Gamma.\mathtt{decrypt}_{\Gamma.sk}(\mu) \wedge \Sigma.\mathtt{verify}_{\Sigma.pk}(\sigma,m) = 0\Big\} \\
\mathit{CS}.\mathtt{convert}(\mu,m) & : \sigma \leftarrow \Gamma.\mathtt{decrypt}_{\Gamma.sk}(\mu)\,;\ b \leftarrow \Sigma.\mathtt{verify}_{\Sigma.pk}(\sigma,m) \\
 & \quad \text{if } b = 0 \text{ return } (\bot) \text{ else return } (\sigma) \\
\mathit{CS}.\mathtt{verifyconverted}(\sigma,m) & : \Sigma.\mathtt{verify}_{\Sigma.pk}(\sigma,m)
\end{array}
$$

Fig. 3.1 The StE paradigm

A CDCS scheme *CS* from the StE paradigm is depicted in Fig. 3.1.

Note that the languages underlying `sconfirm`, `confirm`, and `deny` are in NP and thus accept concurrent zero-knowledge proof systems (Dwork et al. 2004) (Sect. 1.4.2). This guarantees the completeness, soundness, and (offline) non-transferability of the resulting signatures. It further allows the prover (signer or confirmer) to run, *concurrently*, multiple instances of the aforementioned protocols with many verifiers that proceed at times and paces of their choosing; a situation that is likely to occur in the real-life applications of the primitive in question.

3.1.2 Other Variants

The Goldwasser-Waisbard (2004) construction. This construction was the first to circumvent, although partially, the main problem in the basic paradigm, namely the recourse to proofs of general NP statements in the confirmation/denial protocols. The idea consists in considering a class of digital signatures which accept efficient *witness-hiding proofs of knowledge* (WHPoK). A WHPoK (see for instance Goldreich (2001, Sect. 4.6) for more details) is informally a proof where the prover does not reveal the witness but may leak some knowledge during his interaction with the verifier; it is then a weaker notion than zero-knowledge. Let (t, b, s_b) be an accepting transcript resulting from the interaction, between a prover and a verifier, in which the prover convinces the verifier that he holds a digital signature σ on the common input message m. t forms the first message, or the commitment, sent by the prover. $b \xleftarrow{R} \{0, 1\}$ denotes the public coin, or the challenge sent by the verifier. Finally, s_b denotes the response of the prover to the challenge b. It is assumed that given two different accepting transcripts (t, b, s_b) and $(t, 1 - b, s_{1-b})$, there exists a knowledge extractor which can extract the witness, namely the signature σ. With such a class of signatures in addition to an IND-CCA secure encryption scheme Γ, the authors in Goldwasser and Waisbard (2004) provide confirmer signatures on a message m as follows:

1. The signer first produces a digital signature σ on m. Then, he computes the commitment t he would send to the verifier if he wishes to conduct a WHPoK for σ. Next he computes s_0 and s_1, the responses to the challenges $b = 0$ and $b = 1$ resp., along with their encryptions e_0 and e_1 using random coins r_0 and r_1 resp. Finally, the signer sends (t, e_0, e_1) to the signature recipient.
2. The signature recipient selects $b \stackrel{R}{\leftarrow} \{0, 1\}$ and sends it to the signer.
3. The signer reveals s_b to the verifier along with the random coin used to produce its encryption e_b, namely r_b.
4. The signature recipient accepts if e_b is indeed the encryption of s_b using r_b, and if (t, b, s_b) is indeed an accepting transcript for the WHPoK.

The triplet (t, e_0, e_1) forms the confirmer signature the verifier needs to present before the confirmer for verification or conversion. In fact, the confirmer can decrypt both e_0 and e_1 in s_0 and s_1 resp., then extracts the witness σ in case of a valid signature and finally interacts (in case of a confirmation query) with the verifier in a protocol similar to the one above. Conversion is done by revealing σ. And finally, the denial of an invalid signature consists of a ZK proof that the conversion returns an invalid signature.

The construction successfully gets rid of proofs of general NP statements in the confirmation protocol. However, it still resorts to them in the denial protocol. Moreover, the length of the signatures as well as their generation cost grow linearly with the number of rounds in the WHPoK. Finally, the security guarantees satisfied by the construction are much more relaxed compared to those met by the construction realizing the basic StE paradigm. For instance, non-transferability of the signatures may not be guaranteed with the use of WHPoK, as the adversary might get sufficient knowledge (from the confirmation protocol) to convince other parties with the validity of the signature he is holding. Also, the adversary is not given access to a conversion oracle in the non-transferability definition which means that one can say nothing about his ability in transferring knowledge of the validity of signatures when he sees some converted signatures.

The Wikström (2007) construction. This construction does not differ much from the basic StE paradigm in that it consists in first producing a digital signature on the message to be signed then encrypting the resulting signature. The difference is that the used encryption scheme needs to support labels (i.e. tag-based encryption). Actually, the encryption of the digital signatures is done under the tag $\Sigma.pk$, which denotes the public key of the signer. The basic novelty of the work (Wikström 2007) lies in the new security model proposed for confirmer signatures, and in which the construction is analyzed. For instance, security for the confirmer in Wikström (2007), referred to as impersonation resistance, requires that no one should play the role of the genuine confirmer, namely prove that the confirmer key is well formed, that a signature is valid/invalid and finally that a conversion is correct. The formalization of such property is done as usual through a game where the adversary has access to a genuine confirmer oracle that he can consult up to the challenge phase. Consequently, one gets with this definition only a "lunch-time" security for the confirmer unlike the definitions proposed earlier in this text.

3.2 The Exact Unforgeability of StE Constructions

Theorem 3.1 *Given* $(t, q_s) \in \mathbb{N}^2$ *and* $\varepsilon \in [0, 1]$, *confirmer signatures from StE are* (t, ϵ, q_s)-EUF-CMA *secure if and only if the underlying digital signature scheme is* (t, ϵ, q_s)-EUF-CMA *secure.*

Proof Let $\mathcal{A}$ be a (t, ϵ, q_s)-EUF-CMA adversary against the confirmer signatures (from StE). We construct a (t, ϵ, q_s)-EUF-CMA adversary $\mathcal{R}$ against the underlying digital signature scheme as follows.

$\mathcal{R}$ gets the public key of the signature scheme Σ from his challenger. Then he chooses a suitable encryption scheme Γ and gets from $\mathcal{A}$ the generated confirmer key pair $(\Gamma.pk, \Gamma.sk)$.

Signature queries made by $\mathcal{A}$ are answered using $\mathcal{R}$'s challenger to get digital signatures (on the queried messages) which are then encrypted using $\Gamma.pk$ to form the result. Note that $\mathcal{A}$ can check the validity of this signature himself using $\Gamma.sk$.

Eventually, $\mathcal{A}$ outputs a pair (m, μ) consisting of a message m that was never queried for signature and a valid confirmer signature μ on it. $\mathcal{R}$ will simply output $\sigma = \Gamma.\texttt{decrypt}_{\Gamma.sk}(\mu)$ along with m to his own challenger. In fact, σ is a valid digital signature on the message m which was never queried by $\mathcal{R}$ to his own challenger, and thus forms a valid existential forgery on Σ.

Conversely, let (m, σ) be an existential forgery against the digital signature scheme. One can derive a forgery against the confirmer signature by simply encrypting the signature σ using the public key of the confirmer. Simulation of the attacker's environment is easy; the reduction $\mathcal{R}$ (EUF-CMA attacker against the confirmer signature) forwards the appropriate parameters (those concerning the underlying digital signature) to the EUF-CMA attacker $\mathcal{A}$ against the underlying signature scheme. For a signature query on a message m_i, $\mathcal{R}$ first requests his challenger for a confirmer signature μ_i that he decrypts using the private key of the confirmer ($\mathcal{R}$ has access to sk_C) in σ_i, which he outputs to $\mathcal{A}$. At the end, $\mathcal{A}$ outputs a valid digital signature σ on a message m that was never queried for signature. $\mathcal{R}$ encrypts this signature in μ using the public key of the confirmer and outputs the result as a valid existential forgery on m ($\mathcal{R}$ never queried m for a confirmer signature). □

3.2.1 *Roadmap for the Rest of the Chapter*

In the rest of this chapter, we show that StE requires at least IND-PCA secure encryption in order to lead to INV-CMA secure confirmer signatures. We proceed as follows.

First, we rule out the OW-CPA, OW-PCA, and IND-CPA notions by remarking that ElGamal's encryption meets all those notions (under different assumptions), but cannot be used in StE. In fact, the invisibility adversary can create from the challenge signature a new "equivalent" signature (by re-encrypting the ElGamal encryption), and query it for conversion or verification to solve the challenge. Actually, this attack applies to any *homomorphic encryption.*

Next, we show the insufficiency of OW-CCA and NM-CPA secure encryption by means of efficient meta-reductions which forbid the existence of reductions from the invisibility of the resulting confirmer signatures to the OW-CCA or NM-CPA security of the underlying encryption. We first show this impossibility result for a specific kind of reductions, then we extend it to arbitrary reductions assuming further assumptions on the used encryption.

As illustration of our techniques, we provide evidence that the well known Damgård-Pedersen signature (Damgård and Pedersen 1996) is, contrarily to what is conjectured by the authors, unlikely to be indistinguishable under the DDH assumption.

Finally, we show that StE can thrive on IND-PCA encryption provided the used signature scheme is SEUF-CMA secure.

3.3 A Breach in Invisibility with Homomorphic Encryption

Definition 3.1 (Homomorphic Encryption) A homomorphic public-key encryption scheme Γ given by Γ.keygen, Γ.encrypt, and Γ.decrypt has the following properties:

1. The message space $\mathcal{M}$ and the ciphertext space C are groups w.r.t. some binary operations $*_e$ and $\circ_e$ respectively.
2. $\forall (sk, pk) \leftarrow \Gamma.\texttt{keygen}(1^\kappa)$ for any security parameter κ, $\forall m, m' \in \mathcal{M}$:

$$\Gamma.\texttt{encrypt}_{pk}(m *_e m') = \Gamma.\texttt{encrypt}_{pk}(m) \circ_e \Gamma.\texttt{encrypt}_{pk}(m').$$

Examples of homomorphic encryption[1] in the literature include El Gamal (1985), Paillier (1999), and Boneh et al. (2004). All those schemes are IND-CPA secure (under different assumptions).

[1]This encryption is not to confuse with the so-called *fully homomorphic* encryption which preserves the entire ring structure of the plaintexts (supports both addition and multiplication).

In the rest of this text, we refer to such schemes as homomorphic encryption schemes.

Fact 3.1 *The StE paradigm cannot lead to* INV-CMA *secure confirmer signatures when used with homomorphic encryption.*

Proof Let m_0, m_1 be the challenge messages the invisibility adversary $\mathcal{A}$ outputs to his challenger. Let further Γ and Σ denote respectively the homomorphic encryption and the digital signature used as building blocks.

$\mathcal{A}$ gets as a challenge confirmer signature some $\mu_b = \Gamma.\texttt{encrypt}(\Sigma.\texttt{sign}(m_b))$, where $b \xleftarrow{R} \{0,1\}$ and is asked to find b. To solve his challenge, $\mathcal{A}$ obtains another encryption, say $\widetilde{\mu_b}$, of $\Sigma.\texttt{sign}(m_b)$ by multiplying μ_b with an encryption of the identity element. According to the invisibility experiment, $\mathcal{A}$ can query $\widetilde{\mu_b}$ for conversion or verification (w.r.t. either m_0 or m_1) and the answer to such a query is sufficient to conclude. □

Corollary 3.1 *Invisibility in CDCS from StE cannot rest on* OW-CPA, OW-PCA, *or* IND-CPA *secure encryption.*

Proof ElGamal's encryption (El Gamal 1985) is homomorphic and meets the OW-CPA, OW-PCA, or IND-CPA security notions under different assumptions (Paragraph on ElGamal's encryption of Sect. 1.3.1). The rest follows from the previous fact. □

3.4 Impossibility Results for Key-Preserving Reductions

In this section, we show that NM-CPA and OW-CCA secure encryption cannot lead, under a certain type of reductions, to invisible confirmer signatures from StE. We do this by means of efficient *meta-reductions* that use such reductions (the algorithm reducing NM-CPA (OW-CCA) breaking the underlying encryption scheme to breaking the invisibility of the construction) to break the NM-CPA (OW-CCA) security of the encryption scheme. Thus, if the encryption scheme is NM-CPA (OW-CCA) secure, the meta-reductions forbid the existence of such reductions. In case the encryption scheme is not NM-CPA (OW-CCA) secure, such reductions will be useless.

This impossibility result is partial in a first stage since it requires the reduction $\mathcal{R}$, trying to attack a certain property of an encryption scheme given by the public key $\Gamma.pk$, to provide the adversary against the confirmer signature with the confirmer public key $\Gamma.pk$. In other terms, the result applies only to key-preserving reductions (see Definition 1.15). The restriction to such a class of reductions is not unnatural; actually, most if not all the reductions in the literature that base the security of the generic constructions of confirmer signatures on the security of their underlying components, feed the adversary with the public keys of these components (signature schemes, encryption schemes, etc). Next, we use similar techniques to Paillier and Villar (2006) to extend these impossibility results to arbitrary reductions.

3.4.1 *Insufficiency of* OW-CCA *Secure Encryption*

Lemma 3.1 *Assume there exists a key-preserving reduction $\mathcal{R}$ that converts an* INV-CMA *adversary $\mathcal{A}$ against confirmer signatures from the StE paradigm to a* OW-CCA *adversary against the underlying encryption scheme. Then, there exists a meta-reduction $\mathcal{M}$ that* OW-CCA *breaks the encryption scheme in question.*

This lemma claims that if the considered encryption is OW-CCA secure, then, there exists no key-preserving reduction $\mathcal{R}$ that reduces OW-CCA breaking it to INV-CMA breaking the construction, or if there exists such an algorithm, then the underlying encryption is not OW-CCA secure, thus rendering such a reduction useless.

Proof Let $\mathcal{R}$ be the key-preserving reduction that reduces OW-CCA breaking the encryption scheme underlying the construction to INV-CMA breaking the construction itself. We construct an algorithm $\mathcal{M}$ that uses $\mathcal{R}$ to OW-CCA break the same encryption scheme by simulating an execution of the INV-CMA adversary $\mathcal{A}$ against the construction.

Let Γ be the encryption scheme $\mathcal{M}$ is trying to attack w.r.t. key $\Gamma.pk$. $\mathcal{M}$ proceeds as follows.

Let c be the OW-CCA challenge $\mathcal{M}$ is asked to resolve. $\mathcal{M}$ launches $\mathcal{R}$ over Γ under the same key $\Gamma.pk$ and the same challenge c. Obviously, all decryption queries made by $\mathcal{R}$ can be perfectly answered using $\mathcal{M}$'s challenger (since they are different from the challenge c).

$\mathcal{M}$ needs now to simulate an INV-CMA adversary $\mathcal{A}$ to $\mathcal{R}$. To do so, $\mathcal{M}$ picks two random messages m_0 and m_1 from the message space. To ensure that c is not a valid confirmer signature on m_0 or m_1, $\mathcal{M}$ queries $\mathcal{R}$ for the conversion of both (c, m_0) and (c, m_1) and makes sure that the result to both queries is $\perp$. If this is not the case, then $\mathcal{M}$ will simply abort the INV-CMA game, and output the result of the conversion, say σ (which is different from $\perp$), to his own OW-CCA challenger. In fact, with noticeable probability (the probability that the simulation supplied by $\mathcal{R}$ does not deviate from the standard INV-CMA game), σ is a valid decryption of c w.r.t. $\Gamma.pk$.

We assume now that c is not a valid confirmer signature on either m_0 or m_1. Hence, $\mathcal{M}$ outputs m_0, m_1 to $\mathcal{R}$ as challenge messages, and receives a challenge μ_b which is, with noticeable probability (probability that $\mathcal{R}$ supplies a correct simulation), a valid confirmer signature on m_b for $b \in \{0, 1\}$. μ_b is according to our assumption different from the challenge ciphertext c, and $\mathcal{M}$ is requested to find b. To solve his challenge, $\mathcal{M}$ queries his own OW-CCA challenger for the decryption of μ_b. The result to such a query allows $\mathcal{M}$ to find out b with probability one (provided $\mathcal{R}$ supplies a correct simulation).

To sum up, $\mathcal{M}$ is able to perfectly answer the decryption queries made by $\mathcal{R}$ (that are by definition different from the OW-CCA challenge). $\mathcal{M}$ is further capable of successfully simulating an INV-CMA attacker against the construction, provided $\mathcal{R}$ supplies a correct simulation. Thus, $\mathcal{R}$ is expected to return the answer to the OW-CCA challenge. Upon receipt of this answer, $\mathcal{M}$ will forward it to his own challenger. □

3.4.2 *Insufficiency of* NM-CPA *Secure Encryption*

Lemma 3.2 *Assume there exists a key-preserving reduction $\mathcal{R}$ that converts an* INV-CMA *adversary $\mathcal{A}$ against confirmer signatures from the StE paradigm to an* NM-CPA *adversary against the underlying encryption scheme. Then, there exists a meta-reduction $\mathcal{M}$ that* NM-CPA *breaks the used encryption scheme.*

Proof Let $\mathcal{R}$ be the key-preserving reduction that reduces NM-CPA breaking the encryption underlying the construction to INV-CMA breaking the construction itself. We construct an algorithm $\mathcal{M}$ that uses $\mathcal{R}$ to NM-CPA break the same encryption scheme by simulating an execution of the INV-CMA adversary $\mathcal{A}$ against the construction.

Let Γ be the encryption scheme $\mathcal{M}$ is trying to attack w.r.t. public key $\Gamma.pk$. $\mathcal{M}$ will launch $\mathcal{R}$ over the same public key $\Gamma.pk$. Next, $\mathcal{M}$ will simulate an INV-CMA adversary against the constructions:

$\mathcal{M}$ (behaving as $\mathcal{A}$) queries $\mathcal{R}$ on two messages m_0, m_1 ($m_0 \neq m_1$) for confirmer signatures. Let μ_0, μ_1 be the corresponding confirmer signatures resp. $\mathcal{M}$ further queries (μ_i, m_i), $i \in \{0, 1\}$, for conversion. Let σ_0, σ_1 be the corresponding answers respectively. We assume that $\sigma_0 \neq \sigma_1$. If this is not the case, $\mathcal{M}$ repeats the experiment until this holds (if all confirmer signatures are encryptions of the same string σ, then the construction is not secure). At that point, $\mathcal{M}$ outputs $D = \{\sigma_0, \sigma_1\}$ to his NM-CPA challenger as a distribution probability from which the messages will be drawn. He gets a challenge encryption $\mu^\star$, of either σ_0 or σ_1 under $\Gamma.pk$, and is asked to produce a ciphertext μ' whose corresponding plaintext is meaningfully related to the decryption of $\mu^\star$. To solve his task, $\mathcal{M}$ queries for instance $(\mu^\star, m_0)$ for conversion. If the result is σ_0, i.e. $\mu^\star$ is a valid confirmer signature on m_0, then $\mathcal{M}$ outputs $\Gamma.\mathtt{encrypt}_{pk}(\overline{\sigma_0})$ ($\overline{m}$ refers to the bit-complement of m) and the relation R: $R(m, m') = (m' = \overline{m})$. Otherwise, $\mathcal{M}$ outputs $\Gamma.\mathtt{encrypt}_{pk}(\overline{\sigma_1})$ and the same relation R. Finally $\mathcal{M}$ aborts the INV-CMA game.

Clearly, $\mathcal{M}$ correctly solves his NM-CPA challenge if $\mathcal{R}$ provides a correct simulation. □

3.4.3 *Putting All Together*

Theorem 3.2 *Consider the security notions obtained from pairing a security goal* GOAL $\in$ {OW, IND, NM} *and an attack model* ATK $\in$ {CPA, PCA, CCA}. *The encryption scheme underlying the above construction from StE must be at least* IND-PCA *secure, in case the considered reduction is key-preserving, in order to achieve* INV-CMA *secure confirmer signatures.*

Proof Corollary 3.1 rules out OW-CPA, OW-PCA, and IND-CPA encryption. Moreover, Lemma 3.1 and Lemma 3.2 rule out OW-CCA and NM-CPA encryption resp. The next notion to be considered is IND-PCA. □

Remark 3.1 Note that the notions OW-CPA, OW-PCA, and IND-CPA are discarded regardless of the used reduction. In fact, we managed to exhibit an encryption scheme (ElGamal's encryption) which meets all those notions, but leads to insecure confirmer signatures from the StE paradigm.

Remark 3.2 The step of ruling out OW-CPA, OW-PCA, and IND-CPA is necessary although we have proved the insufficiency of stronger notions, namely OW-CCA and NM-CPA. In fact, suppose there is an efficient "useful" key-preserving reduction $\mathcal{R}$ (i.e. $\mathcal{R}$ solves a presumably hard problem) which reduces OW-PCA breaking a cryptosystem Γ underlying a StE construction to INV-CMA breaking the construction itself. Then there exists an efficient key-preserving reduction say $\mathcal{R}'$ that reduces OW-CCA breaking Γ to INV-CMA breaking the construction (OW-CCA is stronger than OW-PCA). This does not contradict Lemma 3.1 as long as Γ is not OW-CCA secure (although it is OW-PCA secure). In other terms, since there are separations between OW-CCA and OW-PCA (same for the other notions), we cannot apply the insufficiency of OW-CCA (NM-CPA) to rule out the weaker notions.

On the Resort to Meta-reductions It is tempting to envisage stronger techniques than meta-reductions in order to achieve the aforementioned negative results. In fact, meta-reductions give only partial results as they consider a specific class of reductions, e.g. key-preserving reductions.

For instance, one might try to adapt existing results that separate security notions in encryption, e.g. Bellare et al. (1998). The problem is that the invisibility adversary in confirmer signatures does not have explicit access to a decryption oracle, i.e. the adversary gets the decryption of a ciphertext only if the latter is a valid confirmer signature on some message (that is also part of the query). Therefore, the separation techniques used in encryption cannot be straightforwardly used in case of confirmer signatures.

Another possibility consists in building simple counter examples of encryption schemes which are OW-CCA (NM-CPA) secure but lead to insecure confirmer signatures when used in the StE paradigm. Again, it seems difficult to achieve results using this approach without assuming special security properties on the used digital signature scheme, i.e. consider signature schemes that are not strongly unforgeable. The merit of meta-reductions lies in achieving separation results regardless of the used digital signature.

We will see in the next section how to extend the obtained negative results if the encryption underlying the constructions satisfies further security properties.

3.5 Extension to Arbitrary Reductions

To extend the results of the previous section to arbitrary reductions, we first define the notion of *non-malleability of an encryption scheme key generator* through the following two games:

In **Game 0**, we consider an algorithm $\mathcal{R}$ trying to break an encryption scheme Γ, w.r.t. a public key $\Gamma.pk$, in the sense of NM-CPA (or OW-CCA) using an adversary $\mathcal{A}$ which solves a problem A, perfectly reducible to OW-CPA breaking the encryption scheme Γ. In this game, $\mathcal{R}$ launches $\mathcal{A}$ over his own challenge key $\Gamma.pk$ and some other parameters chosen freely by $\mathcal{R}$ (according to the specifications of $\mathcal{A}$). We will denote by $\mathsf{Adv}_0(\mathcal{R}^{\mathcal{A}})$ the success probability of $\mathcal{R}$ in such a game, where the probability is taken over the random tapes of both $\mathcal{R}$ and $\mathcal{A}$. We further define $\mathsf{succ}_{\Gamma}^{\mathsf{Game0}}(\mathcal{A}) = \max_{\mathcal{R}} \mathsf{Adv}_0(\mathcal{R}^{\mathcal{A}})$ to be the success in **Game 0** of the best reduction $\mathcal{R}$ making the best possible use of the adversary $\mathcal{A}$. Note that the goal of **Game 0** is to include all key-preserving reductions $\mathcal{R}$ from NM-CPA (or OW-CCA) breaking the encryption scheme in question to solving a problem A, which is reducible to OW-CPA breaking the same encryption scheme.

In **Game 1**, we consider the same entities as in **Game 0**, with the exception of providing $\mathcal{R}$ with, in addition to $\mathcal{A}$, a OW-CPA oracle (i.e. a decryption oracle corresponding to Γ) that he can query w.r.t. any public key $\Gamma.pk' \neq \Gamma.pk$, where $\Gamma.pk$ is the challenge public key of $\mathcal{R}$. Similarly, we define $\mathsf{Adv}_1(\mathcal{R}^{\mathcal{A}})$ to be the success of $\mathcal{R}$ in such a game, and $\mathsf{succ}_{\Gamma}^{\mathsf{Game1}}(\mathcal{A}) = \max_{\mathcal{R}} \mathsf{Adv}_1(\mathcal{R}^{\mathcal{A}})$ the success in **Game 1** of the reduction $\mathcal{R}$ making the best possible use of the adversary $\mathcal{A}$ and of the decryption (OW-CPA) oracle.

Definition 3.2 (Encryption with Non-malleable Key Generator) An encryption scheme Γ is said to have a non-malleable key generator if

$$\Delta = max_{\mathcal{A}} \left| \mathsf{succ}_{\Gamma}^{\mathsf{Game1}}(\mathcal{A}) - \mathsf{succ}_{\Gamma}^{\mathsf{Game0}}(\mathcal{A}) \right|$$

is negligible in the security parameter.

This definition informally means that an encryption scheme has a non-malleable key generator if NM-CPA (or OW-CCA) breaking it w.r.t. a key pk is no easier when given access to a decryption (OW-CPA) oracle w.r.t. any public key $pk' \neq pk$.

We generalize now the previous impossibility results to arbitrary reductions as follows.

Theorem 3.3 *Theorem 3.2 is still valid when considering arbitrary reductions, provided the encryption scheme underlying the constructions has a non-malleable key generator.*

To prove this theorem, we first need the following lemma (similar to Lemma 6 of Paillier and Villar 2006)

Lemma 3.3 *Let $\mathcal{A}$ be an adversary solving a problem A, reducible to* OW-CPA *breaking an encryption scheme Γ, and let $\mathcal{R}$ be an arbitrary reduction $\mathcal{R}$ that* NM-CPA *(*OW-CCA*) breaks Γ given access to $\mathcal{A}$. We have* $\mathsf{Adv}(\mathcal{R}) \leq \mathsf{succ}_{\Gamma}^{\mathsf{Game1}}(\mathcal{A})$.

Proof We will construct an algorithm $\mathcal{M}$ that plays **Game 1** with respect to a perfect oracle for $\mathcal{A}$ and succeeds in breaking the NM-CPA (OW-CCA) security of Γ with the same success probability of $\mathcal{R}$. Algorithm $\mathcal{M}$ gets a challenge w.r.t. a public key *pk* and launches $\mathcal{R}$ over the same challenge and the same public key. If $\mathcal{R}$ calls $\mathcal{A}$ on *pk*, then $\mathcal{M}$ will call his own oracle for $\mathcal{A}$. Otherwise, if $\mathcal{R}$ calls $\mathcal{A}$ on $pk' \neq pk$, $\mathcal{M}$ will invoke his own decryption oracle for pk' (OW-CPA oracle) to answer the queries. In fact, by assumption, the problem A is reducible to OW-CPA breaking Γ. Finally, when $\mathcal{R}$ outputs the result to $\mathcal{M}$, the latter will output the same result to his own challenger. □

The proof of Theorem 3.3 is similar to that of Theorem 5 in Paillier and Villar (2006):

Proof We first remark that the invisibility of StE is perfectly reducible to OW-CPA breaking the encryption scheme underlying the construction. In fact, an invisibility adversary $\mathcal{A}$ on a challenge confirmer signature can first decrypt it, then checks using Σ.`verify` and $\Sigma.pk$ whether the result is a valid digital signature on the message in question or not.

Next, we note that the advantage of the meta-reduction $\mathcal{M}$ in the proof of Lemma 3.2 (Lemma 3.1) is at least the same as the advantage of any key-preserving reduction $\mathcal{R}$ reducing the invisibility of a given confirmer signature to the NM-CPA (OW-CCA) security of its underlying encryption scheme Γ. For instance, this applies to the reduction making the best use of an invisibility adversary $\mathcal{A}$ against the construction. Therefore we have: $\mathsf{succ}_{\Gamma}^{\mathsf{Game0}}(\mathcal{A}) \leq \mathsf{succ}(\mathsf{NM\text{-}CPA}[\Gamma])$, where $\mathsf{succ}(\mathsf{NM\text{-}CPA}[\Gamma])$ is the success of breaking Γ in the NM-CPA sense. We also have $\mathsf{succ}_{\Gamma}^{\mathsf{Game0}}(\mathcal{A}) \leq \mathsf{succ}(\mathsf{OW\text{-}CCA}[\Gamma])$. Now, Let $\mathcal{R}$ be an arbitrary reduction from NM-CPA (OW-CCA) breaking an encryption scheme Γ, with a non-malleable key generator, to INV-CMA breaking the construction (using the same encryption scheme Γ). We have

$$\begin{aligned}
\mathsf{Adv}(\mathcal{R}) &\leq \mathsf{succ}_{\Gamma}^{\mathsf{Game1}}(\mathcal{A}) \\
&\leq \mathsf{succ}_{\Gamma}^{\mathsf{Game0}}(\mathcal{A}) + \Delta \\
&\leq \min\{\mathsf{succ}(\mathsf{NM\text{-}CPA}[\Gamma], \mathsf{succ}(\mathsf{OW\text{-}CCA}[\Gamma])\} + \Delta
\end{aligned}$$

since Δ is negligible, if Γ is NM-CPA (OW-CCA) secure, then the advantage of $\mathcal{R}$ is also negligible. □

Existence of Encryption with a Non-malleable Key Generator It is not difficult to see that factoring or RSA based encryption schemes are the first candidates to have a non-malleable key generator. In fact, if the public key in these schemes consists only of an RSA modulus *n*, then factorization of other moduli will not help factoring *n*. Examples of such schemes are countless and include OAEP (Bellare and Rogaway 1993), REACT-RSA (Okamoto and Pointcheval 2001), PKCS#1 v2.2 (PKC 2012), Rabin and related systems (Blum and Goldwasser 1984; Chor and Goldreich 1984; Williams 1980), the EPOC family, Paillier (1999), etc.

Discrete-log-based encryption schemes fail however into this category. Actually, a discrete-log oracle w.r.t. some generator of a given group is sufficient to extract the discrete-log of any element (w.r.t. any element) in this group. Therefore, extension of the above separation results is not straightforward for these schemes; it must use the specific properties of the used encryption scheme. We provide an illustration of such an extension in the subsequent section.

3.6 Analysis of Damgård-Pedersen's Undeniable Signatures

In this section, we analyze the invisibility of Damgård-Pedersen's undeniable signature; recall that undeniable signatures are confirmer signatures where the signer and the confirmer are one entity. We provide evidence that this signature is unlikely to meet its conjectured security.

Let $m \in \{0,1\}^\star$ be an arbitrary message. Damgård-Pedersen's undeniable signature consists of the following procedures:

$\mathtt{setup}(\mathbf{1}^\kappa)$ On input the security parameter κ, generate a k-bit prime t and a prime $p \equiv 1 \bmod t$. Furthermore, select a collision-resistant hash function H that maps arbitrary-length messages to $\mathbb{Z}_t$.

$\mathtt{keygen}(\mathbf{1}^\kappa)$ Generate g of order t, $x \in \mathbb{Z}_t^\times$, and $h \equiv g^x \bmod p$. Furthermore, select a generator α of $\mathbb{Z}_t^\times$ and $\nu \in \{0,1,\ldots,t-1\}$, and compute $\beta \equiv \alpha^\nu \bmod t$. The public key is $pk = (p,t,g,h,\alpha,\beta)$ and the private key is $sk = (x,\nu)$.

$\mathtt{sign}_{sk}(\mathbf{m})$ The signer first computes an ElGamal signature (s,r) on m, i.e. compute $r \equiv g^b \bmod p$ for some $b \xleftarrow{R} \mathbb{Z}_t^\times$, then compute s as $h(m) \equiv rx + bs \bmod t$. Next, he computes an ElGamal encryption $(E_1 = \alpha^\rho, E_2 = s\beta^\rho) \bmod t$, for $\rho \xleftarrow{R} \mathbb{Z}_{t-1}$, of s. The undeniable signature on m is the triple (E_1,E_2,r).

$\{\mathtt{confirm},\mathtt{deny}\}([\mathbf{E_1},\mathbf{E_2},\mathbf{r}],\mathbf{m},pk)$ To confirm (deny) a purported signature (E_1,E_2,r) on a certain message m, the signer issues the following proof (see Damgård and Pedersen 1996)

$$\mathsf{ZKPoK}\left\{s\colon \mathsf{DL}_\alpha(\beta) = \mathsf{DL}_{E_1}(E_2 \cdot s^{-1}) \wedge g^{h(m)}h^{-r} = (\neq) r^s\right\}$$

In Damgård and Pedersen (1996), the authors prove that the above signatures are unforgeable if the underlying ElGamal signature is also unforgeable, and they conjecture that the signatures meet the following invisibility notion if the problem DDH is hard:

Definition 3.3 (Indistinguishability in Damgård-Pedersen's Signatures) It is defined through the following game between an attacker $\mathcal{A}$ (*a distinguisher*) and his challenger $\mathcal{R}$.

Phase 1 after $\mathcal{A}$ gets the public key, pk, of the scheme from $\mathcal{R}$, he starts issuing *status requests* and *signature requests*. In a status request, $\mathcal{A}$ produces a pair (m,z), and receives a 1-bit answer which is 1 iff z is a valid undeniable signature on m w.r.t. pk. In a signature request, $\mathcal{A}$ produces a message m and receives an undeniable signature z on it w.r.t. pk.

Challenge Once $\mathcal{A}$ decides that **Phase 1** is over, he outputs a message m and receives a string z which is either a valid undeniable signature on m (w.r.t. pk) or a randomly chosen string from the signature space.

Phase 2 $\mathcal{A}$ resumes adaptively making the previous types of queries, provided that m does not occur in any request, and that z does not occur in any status request. Eventually, $\mathcal{A}$ will output a bit.

Let p_r, resp. p_s be the probability that $\mathcal{A}$ answers 1 in the real, resp. the simulated case. Both probabilities are taken over the random coins of both $\mathcal{A}$ and $\mathcal{R}$. We say that the signatures are indistinguishable if $|p_r - p_s|$ is a negligible function in the security parameter.

It is clear that the Damgård-Pedersen signatures do not provide the INV-CMA notion according to Sect. 3.3. In the rest of this section, we provide evidence that the Damgård-Pedersen signatures are unlikely to meet the above indistinguishability notion under the DDH assumption.

Lemma 3.4 *Assume there exists a key-preserving reduction $\mathcal{R}$ that uses an indistinguishability adversary $\mathcal{A}$ (in the sense of Definition 3.3) against the above scheme to solve the DDH problem. Then, there exists an efficient meta-reduction $\mathcal{M}$ that solves the DDH problem.*

Proof Let $\mathcal{R}$ be the key-preserving reduction that reduces the DDH problem to distinguishing the Damgård-Pedersen signatures in the sense of Definition 3.3. We will construct an algorithm $\mathcal{M}$ that uses $\mathcal{R}$ to solve the DDH problem by simulating a distinguisher against the signatures.

Let $(\alpha, \beta, c_1 = \alpha^a, c_2 = \beta^b) \in \mathbb{Z}_t^\times \times \mathbb{Z}_t^\times$ be the DDH instance $\mathcal{M}$ is asked to solve, that is, $\mathcal{M}$ is asked to say whether $a \equiv b \bmod (t-1)$ or not. $\mathcal{M}$ acting as a distinguisher of the signature will make a signature request on an arbitrary message m. Let (E_1, E_2, r) be the answer to such a query. $\mathcal{M}$ will make now a status query on $(c_1 \cdot E_1, c_2 \cdot E_2, r)$ and the message m. $(\alpha, \beta, c_1, c_2)$ is a yes-Diffie-Hellman instance ($a \equiv b \bmod (t-1)$) iff the result of the last query is the confirmation that $(c_1 \cdot E_1, c_2 \cdot E_2, r)$ is a signature on m. □

Extension to Arbitrary Reductions We cannot employ, in this case, the non-malleability of the key generator technique discussed above. In fact, this would correspond to assuming that the DDH problem, w.r.t. a given public key pk, is difficult even when given access to a CDH oracle w.r.t. any $pk' \neq pk$, which is untrue.

However, we can see that the result still holds true if $\mathcal{R}$ feeds the adversary $\mathcal{A}$ with the confirmer key $(\alpha', \beta') = (\alpha^\ell, \beta^\ell)$ for some ℓ not necessarily known to $\mathcal{M}$ (recall that DDH is random self-reducible according to Remark 1.3). In fact, $\mathcal{M}$ (or $\mathcal{A}$) checks that $(\alpha', \beta', \alpha, \beta)$ is a Diffie-Hellman tuple (i.e. a yes Diffie-Hellman instance) by first making a signature request on some message, then making a status request on the same message and on the product of the corresponding confirmer signature and the tuple $(\alpha, \beta, 1)$ (the answer to such a status request should be the execution of the confirmation oracle). Next, $\mathcal{A}$ checks his DDH instance $(\alpha, \beta, c_1, c_2)$ by using the same technique, namely first make a signature request on some message, followed by a status request on the same message and the product of the returned signature with the tuple $(c_1, c_2, 1)$. The answer to this query is sufficient for $\mathcal{M}$ to conclude.

Finally, Damgård-Pedersen's undeniable signatures can be repaired so as to provide invisibility by producing an ElGamal signature on the message to be signed concatenated with the used encapsulation E_1. In the following chapter, we will see that this repair turns out to be special instance of the new proposed StE paradigm for confirmer signatures.

3.7 Sufficiency of IND-PCA Secure Encryption

The above negative results are due to the *strong forgeability* of confirmer signatures from StE. Actually, a polynomial-time adversary is able to produce, given a valid confirmer signature on some message, another valid confirmer signature on the same message without the help of the signer; the attacker requests the conversion of the given confirmer signature and then obtains a new confirmer signature on the same message by simply re-encrypting the response (note that a conversion query is not necessary if the used encryption scheme is homomorphic according to Sect. 3.3). Therefore, any reduction $\mathcal{R}$ from the invisibility of the construction to the security of the underlying encryption scheme will need more than a list of records maintaining the queried messages along with the corresponding confirmer and digital signatures. Thus the insufficiency of notions like IND-CPA. In Camenisch and Michels (2000), the authors stipulate that the given reduction would need a decryption oracle (of the encryption scheme) in order to handle the queries made by the INV-CMA attacker $\mathcal{A}$, which makes the invisibility of the constructions rest on the IND-CCA security of the encryption scheme. In this section, we remark that the queries made by $\mathcal{A}$ are not completely uncontrolled by $\mathcal{R}$; they are encryptions of some data already released by $\mathcal{R}$, provided the digital signature scheme is strongly unforgeable, and thus known to her. Therefore, a plaintext-checking oracle suffices to handle those queries.

Theorem 3.4 (Invisibility of StE) *Given $(t, q_s, q_v, q_{sc}) \in \mathbb{N}^4$ and $(\epsilon, \epsilon') \in [0, 1]^2$, confirmer signatures from StE are $(t, \epsilon, q_s, q_v, q_{sc})$-INV-CMA secure if the underlying digital signature is (t, ϵ', q_s)-SEUF-CMA secure and the underlying encryption scheme is $(t + q_s q_{sc}(q_{sc} + q_v), \epsilon \cdot (1 - \epsilon')^{(q_{sc}+q_v)}, q_{sc}(q_{sc} + q_v))$-IND-PCA secure.*

Proof Let $\mathcal{A}$ be an attacker that $(t, \epsilon, q_s, q_v, q_{sc})$-INV-CMA breaks a confirmer signature from StE, believed to use a (t, ϵ', q_s)-SEUF-CMA signature scheme. We construct an algorithm $\mathcal{R}$ that IND-PCA breaks the underlying encryption.

[keygen] $\mathcal{R}$ gets the public parameters of the target encryption scheme from her challenger, that are $\Gamma.pk$, $\Gamma.$`encrypt`, and $\Gamma.$`decrypt`. Then, she chooses a secure signature scheme Σ with parameters $\Sigma.pk$, $\Sigma.sk$, $\Sigma.$`sign`, and $\Sigma.$`verify`.

[`sign` **queries**] For a signature query on a message m, $\mathcal{R}$ proceeds exactly as the standard algorithm using $\Sigma.sk$ and $\Gamma.pk$. $\mathcal{R}$ further maintains *internally* in a list $\mathcal{L}$ the queried messages along with the corresponding confirmer signatures and the intermediate values namely the digital signatures (on the message) and the random values used to produce the confirmer signatures. It is clear that this simulation is indistinguishable from the standard `sign` algorithm.

[`sconfirm` **queries**] $\mathcal{R}$ executes the standard `sconfirm` protocol on a just generated signature using the randomness used to produce the confirmer signature in question.

[`convert` **queries**] For a putative confirmer signature μ on m, $\mathcal{R}$ will look up the list $\mathcal{L}$. We note that each record of $\mathcal{L}$ comprises four components, namely, (1) m_i : the queried message (2) σ_i : the digital signature on m_i (3) $\mu_i = \Gamma.\texttt{encrypt}_{\Gamma.pk}(\sigma_i)$: the confirmer signature on m_i (4) r_i : the randomness used to encrypt σ_i in μ_i.

If no record having as first component the message m appears in $\mathcal{L}$, then $\mathcal{R}$ will output $\perp$.

Otherwise, let t be the number of records having as first component the message m. $\mathcal{R}$ will invoke the plaintext-checking oracle (PCA) furnished by her own challenger on each (σ_i, μ), where σ_i corresponds to the second component of such records (the PCA oracle is invoked t times). If the PCA oracle identifies μ as a valid encryption of some σ_i, then $\mathcal{R}$ will return σ_i, otherwise she will return $\perp$.

This simulation differs from the real one when the signature μ is valid and was not obtained from the signing oracle. We note that the only ways to create a valid confirmer signature without the help of $\mathcal{R}$ consist in either encrypting a digital signature obtained from the conversion oracle or coming up with a new fresh pair of message and corresponding signature (m, μ). $\mathcal{R}$ can handle the first case using her PCA oracle and list of records $\mathcal{L}$. In the second case, we can distinguish two sub-cases: either m has not been queried to the signing oracle in which case the pair (m, μ) corresponds to an existential forgery on the confirmer signature scheme and thus to an existential forgery on the underlying digital scheme according to Theorem 3.1, or m has been queried to the signing oracle but $\Gamma.\texttt{decrypt}(\mu)$ is not an output of the selective conversion oracle, which corresponds to a strong existential forgery on the underlying digital signature. Therefore, the probability that this scenario does not happen is at least $(1 - \epsilon')^{q_{sc}}$ because the underlying digital signature scheme is (t, ϵ', q_s)-SEUF-CMA secure by assumption.

[{`confirm`, `deny`} **queries**] $\mathcal{R}$ will proceed exactly as in the selective conversion with the exception of simulating the denial protocol instead of returning $\perp$, or the confirmation protocol instead of returning the converted digital signature (the {`confirm`, `deny`} protocols are concurrent zero-knowledge proofs, and thus they are simulatable). This simulation does not deviate from the standard execution of the protocols by at least $(1 - \epsilon')^{q_v}$.

[Challenge phase] Eventually, $\mathcal{A}$ outputs two challenge messages m_0 and m_1. $\mathcal{R}$ will then compute two signatures σ_0 and σ_1 on m_0 and m_1 respectively, which she gives to her own IND-PCA challenger. $\mathcal{R}$ receives then the challenge $\mu^\star$, as the encryption of either σ_0 or σ_1, which she will forward to $\mathcal{A}$.

[Post challenge phase] $\mathcal{A}$ will continue issuing queries to the signing, confirmation/denial and selective conversion oracles and $\mathcal{R}$ can answer as previously. Note that in this phase, $\mathcal{A}$ is not allowed to query the selective conversion or the confirmation/denial oracles on $(m_i, \mu^\star)$, $i = 0, 1$. Also, $\mathcal{R}$ is not allowed to query her PCA oracle on $(\mu^\star, \sigma_i)$, $i = 0, 1$. If during the selective conversion or confirmation/denial queries made by $\mathcal{A}$, $\mathcal{R}$ is compelled to query her PCA oracle on $(\mu^\star, \sigma_i)$, $i = 0, 1$, she will simply output $\perp$ in case of a selective conversion query or simulate the denial protocol in case of a verification query. This differs from the real scenario when $\mu^\star$ is a valid confirmer signature on some message $m \notin \{m_0, m_1\}$, which corresponds to an existential forgery on the underlying signature scheme (σ_i will be a valid digital signature on m_0 or m_1 and on a message $m \notin \{m_0, m_1\}$). Again, the probability to not deviate from the real invisibility game is at least $(1 - \epsilon')^{q_{sc}+q_v}$.

[Final output] When $\mathcal{A}$ outputs his answer $b \in \{0, 1\}$, $\mathcal{R}$ will forward this very answer to her own challenger. Therefore $\mathcal{R}$ will IND-PCA break the underlying encryption scheme with advantage at least $\epsilon \cdot (1 - \epsilon')^{(q_v+q_{sc})}$, in time at most $t + q_s q_{sc}(q_v + q_{sc})$ after at most $q_{sc}(q_{sc} + q_v)$ queries to the PCA oracle.

□

References

Bellare M, Rogaway P (1993) Random Oracles are practical: a paradigm for designing efficient protocols. In: Denning D, Pyle R, Ganesan R, Sandhu R, Ashby V (eds) Proceedings of the first ACM conference on computer and communications security. ACM Press, New York, pp 62–73

Bellare M, Desai A, Pointcheval D, Rogaway P (1998) Relations among notions of security for public-key encryption schemes. In: Krawczyk H (ed) Advances in cryptology - CRYPTO'98. LNCS, vol 1462. Springer, Heidelberg, pp 26–45

Blum M, Goldwasser S (1984) An efficient probabilistic public-key encryption scheme which hides all partial information. In: Proceedings of advances in cryptology, proceedings of CRYPTO '84, Santa Barbara, CA, 19–22 August 1984, pp 289–302. http://dx.doi.org/10.1007/3-540-39568-7_23

Boneh D, Boyen X, Shacham H (2004) Short group signatures. In: Franklin MK (ed) Advances in cryptology - CRYPTO 2004. LNCS, vol 3152. Springer, Heidelberg, pp 41–55

Camenisch J, Michels M (2000) Confirmer signature schemes secure against adaptative adversaries. In: Preneel B (ed) Advances in cryptology - EUROCRYPT 2000. LNCS, vol 1807. Springer, Heidelberg, pp 243–258

Chor B, Goldreich O (1984) RSA/Rabin least significant bits are 1/2 + 1/(poly(log N)) secure. In: Blakley GR, Chaum D (eds) Proceedings of advances in cryptology, proceedings of CRYPTO '84, Santa Barbara, CA, 19–22 August 1984. LNCS, vol 196. Springer, Heidelberg, pp 303–313. http://dx.doi.org/10.1007/3-540-39568-7_24

Damgård IB, Pedersen TP (1996) New convertible undeniable signature schemes. In: Maurer UM (ed) Advances in cryptology - EUROCRYPT'96. LNCS, vol 1070. Springer, Heidelberg, pp 372–386

Dwork C, Naor M, Sahai A (2004) Concurrent zero-knowledge. J Assoc Comput Mach 51(6):851–898

El Gamal T (1985) A public key cryptosystem and a signature scheme based on discrete logarithms. IEEE Trans Inf Theory 31:469–472

Goldreich O (2001) Foundations of cryptography. Basic tools. Cambridge University Press, Cambridge

Goldwasser S, Waisbard E (2004) Transformation of digital signature schemes into designated confirmer signature schemes. In: Naor M (ed) Theory of cryptography, TCC 2004. LNCS, vol 2951. Springer, Heidelberg, pp 77–100

Okamoto T, Pointcheval D (2001) The gap-problems: a new class of problems for the security of cryptographic schemes. In: Kim K (ed) 4th International workshop on practice and theory in public key cryptography, PKC 2001. LNCS, vol 1992. Springer, Heidelberg, pp 104–118

Paillier P (1999) Public-key cryptosystems based on composite degree residuosity classes. In: Stern J (ed) EUROCRYPT. LNCS, vol 1592. Springer, Heidelberg, pp 223–238

Paillier P, Villar J (2006) Trading one-wayness against chosen-ciphertext security in factoring-based encryption. In: Lai X, Chen K (eds) ASIACRYPT. LNCS, vol 4284. Springer, Heidelberg, pp 252–266

PKC (2012) PKCS #1 v2.2: RSA cryptography standard. RSA Laboratories. http://www.emc.com/collateral/white-papers/h11300-pkcs-1v2-2-rsa-cryptography-standard-wp.pdf

Wikström D (2007) Designated confirmer signatures revisited. In: Vadhan SP (ed) TCC 2007. LNCS, vol 4392. Springer, Heidelberg, pp 342–361

Williams HC (1980) A modification of the RSA public-key encryption procedure (Corresp.). IEEE Trans Inf Theory 26(6):726–729. http://doi.ieeecomputersociety.org/10.1109/TIT.1980.1056264

Chapter 4
An Efficient Variant of StE

Abstract The study conducted in the previous chapter concludes that the basic StE paradigm imposes IND-PCA secure encryption in order to reach invisibility. This condition on the base encryption excludes a class of encryption schemes that allows for a great efficiency of the confirmation/denial protocols. In this chapter, we propose an effective variation of StE; we demonstrate its efficiency by explicitly describing the confirmation/denial protocols when the building blocks are instantiated from a large class of signature/encryption schemes. The modification we propose applies only to the confirmer signature case; we refer to Chap. 7 for an alternative paradigm for verifiable signcryption.

4.1 The New StE

One attempt to circumvent the problem of *strong forgeability* of constructions obtained from the plain StE paradigm can be achieved by binding the digital signature to its encryption. In this way, from a digital signature σ and a message m, an adversary cannot create a new confirmer signature on m by just re-encrypting σ. In fact, σ forms a digital signature on m and some data, say c, which uniquely defines the confirmer signature on m. Moreover, this data c has to be public in order to issue the $\{\texttt{sconfirm}, \texttt{confirm}, \texttt{deny}\}$ protocols.

In this section, we propose a realization of this idea using hybrid encryption (the KEM/DEM paradigm). We also allow more flexibility without compromising the overall security by encrypting only one part of the signature and leaving out the other part, provided it does not reveal information about the key or the message.

4.1.1 Construction

Consider the following components

- **A digital signature scheme** Σ given by: (1) $\Sigma.\texttt{keygen}$ which generates a key pair ($\Sigma.sk$, $\Sigma.pk$) (2) $\Sigma.\texttt{sign}$ (3) $\Sigma.\texttt{verify}$.

L. El Aimani, *Verifiable Composition of Signature and Encryption*,
https://doi.org/10.1007/978-3-319-68112-2_4

- **A KEM** $\mathcal{K}$ given by: (1) $\mathcal{K}$.keygen which generates a key pair ($\mathcal{K}.pk$, $\mathcal{K}.sk$) (2) $\mathcal{K}$.encap (3) $\mathcal{K}$.decap.
- **A DEM** $\mathcal{D}$ given by: (1) $\mathcal{D}$.encrypt (2) $\mathcal{D}$.decrypt.

Assumption on Σ We assume that any digital signature σ, generated using Σ on an arbitrary message m, can be efficiently transformed in a reversible way to a pair (s, r) where r reveals no information about m nor about ($\Sigma.sk$, $\Sigma.pk$). In other words, there exists an algorithm that inputs a message m and a key pair ($\Sigma.sk$, $\Sigma.pk$) and outputs a string statistically indistinguishable from r, where the probability is taken over the messages and the key pairs considered by Σ. This technical detail will improve the efficiency of the construction as it will not necessitate encrypting the entire signature σ, but only the message-key-dependent part, namely s. Finally, we assume that s belongs to the message space of $\mathcal{D}$.

Assumption on $\mathcal{K}$ We further assume that the encapsulations generated by $\mathcal{K}$ are exactly κ-bit long, where κ is a security parameter. This can be for example realized by padding with zeros, on the left of the most significant bit of the given encapsulation, until the resulting string has length κ. Moreover, the operator $\|$ denotes the usual concatenation operation between two bit-strings. As a result, the first bit of m will always be at the $(\kappa + 1)$th position in $c\|m$, where c is a given encapsulation.

The construction of confirmer signatures from Σ, $\mathcal{K}$, and $\mathcal{D}$ is given in Fig. 4.1. Recall that $(\mathcal{K}, \mathcal{D})$ refers to the public-key encryption scheme resulting from the combination of the KEM $\mathcal{K}$ and the DEM $\mathcal{D}$ using the hybrid encryption paradigm.

$$
\begin{array}{|ll|}
\hline
\texttt{CS.setup}(1^\kappa) & : \Sigma.\texttt{setup}(1^\kappa)\,;\ \mathcal{K}.\texttt{setup}(1^\kappa)\,;\ \mathcal{D}.\texttt{setup}(1^\kappa) \\
\texttt{CS.keygen}_S(1^\kappa) & : \Sigma.\texttt{keygen}(1^\kappa) \\
\texttt{CS.keygen}_C(1^\kappa) & : \mathcal{K}.\texttt{keygen}(1^\kappa) \\
\texttt{CS.sign}(m) & : (c,k) \leftarrow \mathcal{K}.\texttt{encap}_{\{\mathcal{K}.pk,coins\}}()\,;\ (s,r) \leftarrow \Sigma.\texttt{sign}_{\Sigma.sk}(c\|m) \\
 & \quad \text{return}(c, \mathcal{D}.\texttt{encrypt}_k(s), r) \\
\texttt{CS.sconfirm}([c,e,r],m) & : \text{ZKP}\big\{(s,coins) : (c,e) = (\mathcal{K},\mathcal{D}).\texttt{encrypt}_{\{\mathcal{K}.pk,coins\}}(s) \ \wedge \\
 & \qquad \Sigma.\texttt{verify}_{\Sigma.pk}([s,r],c\|m) = 1 \big\} \\
\texttt{CS.confirm}([c,e,r],m) & : \text{ZKP}\big\{(s,\mathcal{K}.sk) : s = (\mathcal{K},\mathcal{D}).\texttt{decrypt}_{\{\mathcal{K}.sk\}}(c,e) \ \wedge \\
 & \qquad \Sigma.\texttt{verify}_{\Sigma.pk}([s,r],c\|m) = 1 \big\} \\
\texttt{CS.deny}([c,e,r],m) & : \text{ZKP}\big\{(s,\mathcal{K}.sk) : s = (\mathcal{K},\mathcal{D}).\texttt{decrypt}_{\{\mathcal{K}.sk\}}(c,e) \ \wedge \\
 & \qquad \Sigma.\texttt{verify}_{\Sigma.pk}([s,r],c\|m) = 0 \big\} \\
\texttt{CS.convert}([c,e,r],m) & : s \leftarrow (\mathcal{K},\mathcal{D}).\texttt{decrypt}_{\mathcal{K}.sk}(c,e)\,;\ b \leftarrow \Sigma.\texttt{verify}_{\Sigma.pk}([s,r],c\|m) \\
 & \quad \text{if } b = 0 \text{ return } (\perp) \text{ else return } (s,r,c) \\
\hline
\end{array}
$$

Fig. 4.1 The new StE paradigm

4.1.2 Security Analysis

Theorem 4.1 (Unforgeability of the New StE) *Given* $(t, q_s) \in \mathbb{N}^2$ *and* $\varepsilon \in [0, 1]$, *the above construction is* (t, ϵ, q_s)-EUF-CMA *secure if the underlying digital signature scheme is* (t, ϵ, q_s)-EUF-CMA *secure.*

Proof Let $\mathcal{A}$ be an attacker that (t, ϵ, q_s)-EUF-CMA breaks the above construction. The algorithm $\mathcal{R}$ (t, ϵ, q_s)-EUF-CMA breaks the underlying digital signature scheme Σ as follows:

[keygen] $\mathcal{R}$ gets the parameters of Σ from her challenger. Then, she chooses appropriate KEM $\mathcal{K}$ and DEM $\mathcal{D}$ and asks $\mathcal{A}$ to provide her with the confirmer key pair $(\mathcal{K}.sk, \mathcal{K}.pk)$. Finally, $\mathcal{R}$ fixes the above parameters as a setting for the confirmer signature scheme $\mathcal{A}$ is trying to attack.

[sign **queries**] For a signature query on a message m, $\mathcal{R}$ computes an encapsulation c together with its decapsulation k (using $\Gamma.pk$). Then, she requests her challenger for a digital signature $\sigma = (s, r)$ on $c \| m$. Finally, she encrypts s in $\mathcal{D}.\mathtt{encrypt}_k(s)$ and outputs the confirmer signature $(c, \mathcal{D}.\mathtt{encrypt}_k(s), r)$.

[Final Output] Once $\mathcal{A}$ outputs his forgery $\mu^\star = (c^\star, e^\star, r^\star)$ on $m^\star$. $\mathcal{R}$ will compute the decapsulation of $c^\star$, say $k^\star$. If $\mu^\star$ is valid then by definition $(\mathcal{D}.\mathtt{decrypt}_{k^\star}(e^\star), r^\star)$ is a valid digital signature on $c^\star \| m^\star$. Thus, $\mathcal{R}$ outputs $(\mathcal{D}.\mathtt{decrypt}_{k^\star}(e^\star), r^\star)$ and $c^\star \| m^\star$ as a valid existential forgery on Σ. In fact, if, during a query made by $\mathcal{A}$ on a message m_i, $\mathcal{R}$ is compelled to query her own challenger for a digital signature on $c^\star \| m^\star = c_i \| m_i$, then $m^\star = m_i$ (by construction), which contradicts the fact that $(\mu^\star, m^\star)$ is an existential forgery output by $\mathcal{A}$.

□

The following remark is vital for the invisibility of the resulting confirmer signatures.

Remark 4.1 (Strong Unforgeability of the New StE) The previous theorem shows that existential unforgeability of the underlying digital signature scheme suffices to ensure existential unforgeability of the resulting construction. Actually, one can also show that this requirement on the digital signature (EUF-CMA security) guarantees that no adversary, against the construction, can come up with a valid confirmer signature $\mu = (c, e, r)$ (c is the encapsulation used to generate the confirmer signature μ) on a message m that has been queried before to the signing oracle but where c was never used to generate answers (confirmer signatures) to the signature queries.

To prove this claim, we construct from such an adversary, say $\mathcal{A}$, an EUF-CMA adversary $\mathcal{R}$ against the underlying digital signature scheme, which runs in the same time and has the same advantage as $\mathcal{A}$. In fact, $\mathcal{R}$ will simulate $\mathcal{A}$'s environment in the same way described in the proof of Theorem 4.1. When $\mathcal{A}$ outputs his forgery $\mu^\star = (c^\star, e^\star, r^\star)$ on a message m_i that has been previously queried to the signing oracle, $\mathcal{R}$ decrypts $(c^\star, e^\star)$ in $s^\star$, which by definition forms, together with $r^\star$, a valid digital signature on $c^\star \| m_i$. Since by assumption $c^\star$ was never used to generate confirmer signatures on the queried messages, $\mathcal{R}$ never invoked her own challenger for a digital signature on $c^\star \| m_i$. Therefore, $(s^\star, r^\star)$ along with $c^\star \| m_i$ will form a valid existential forgery on the underlying digital signature scheme.

In the rest of this section, we show that the new StE paradigm achieves a stronger notion (than INV-CMA) of invisibility that we denote SINV-CMA. This notion was first introduced in Galbraith and Mao (2003), and it captures the difficulty to distinguish confirmer signatures on an adversarially chosen message from random elements in the confirmer signature space. The difference with the previously mentioned notion lies in the challenge phase where the SINV-CMA attacker outputs a message $m^\star$ and receives in return an element $\mu^\star$ which is either a confirmer signature on $m^\star$ or a random element from the confirmer signature space. There is again the natural restriction of not querying the challenge pair to the `sconfirm`, {`confirm`, `deny`}, and `convert` oracles.

Note that this stronger notion of invisibility captures both the traditional invisibility INV-CMA and the anonymity of the confirmer signatures, i.e. the difficulty to distinguish confirmer signatures based on the keys under which they are created. We refer to Galbraith and Mao (2003) for the details.

Definition 4.1 (Strong Invisibility for Confirmer Signatures) Let *CS* be a CDCS scheme and $\mathcal{A}$ be a PPTM. We consider the following experiment where κ is a security parameter.

Experiment $\mathbf{Exp}_{CS,\mathcal{A}}^{\mathsf{SINV\text{-}CMA}}(1^\kappa)$

1. $param \leftarrow CS.\mathtt{setup}(1^\kappa)$
2. $(pk_{\mathsf{S}}, sk_{\mathsf{S}}) \leftarrow CS.\mathtt{keygen}_{\mathsf{S}}(1^\kappa)$; $(pk_{\mathsf{C}}, sk_{\mathsf{C}}) \leftarrow CS.\mathtt{keygen}_{\mathsf{C}}(1^\kappa)$
3. $(m^\star, \mathcal{I}) \leftarrow \mathcal{A}^{\mathfrak{S},\mathfrak{Cv},\mathfrak{V}}(\mathsf{find}, pk_{\mathsf{S}}, pk_{\mathsf{C}})$
   - $\mathfrak{S} : m \longmapsto CS.\mathtt{sign}_{sk_{\mathsf{S}}}(m, pk_{\mathsf{C}})$
   - $\mathfrak{Cv} : (\mu, m) \longmapsto CS.\mathtt{convert}_{sk_{\mathsf{C}}}(\mu, m)$
   - $\mathfrak{V} : (\mu, m) \longmapsto CS.\{\mathtt{sconfirm}, \mathtt{confirm}, \mathtt{deny}\}(\mu, m)$
4. $\mu_1^\star \leftarrow CS.\mathtt{sign}_{sk_{\mathsf{S}}}(m^\star, pk_{\mathsf{C}})$; $\mu_0^\star \xleftarrow{R} CS.\mathsf{space}$
5. $b \xleftarrow{R} \{0, 1\}$
6. $b^\star \leftarrow \mathcal{A}^{\mathfrak{S},\mathfrak{Cv},\mathfrak{V}}(\mathsf{guess}, \mathcal{I}, \mu_b^\star, pk_{\mathsf{S}}, pk_{\mathsf{C}})$
   - $\mathfrak{S} : m \longmapsto CS.\mathtt{sign}_{\{sk_{\mathsf{S}}, pk_{\mathsf{C}}\}}(m)$
   - $\mathfrak{Cv} : (\mu, m)(\neq (\mu_b^\star, m^\star)) \longmapsto CS.\mathtt{convert}_{sk_{\mathsf{C}}}(\mu, m)$
   - $\mathfrak{V} : (\mu, m)(\neq (\mu_b^\star, m^\star)) \longmapsto CS.\{\mathtt{sconfirm}, \mathtt{confirm}, \mathtt{deny}\}(\mu, m)$
7. return $(b = b^\star)$

We define $\mathcal{A}$'s advantage as

$$\mathsf{Adv}_{CS,\mathcal{A}}^{\mathsf{SINV\text{-}CMA}}(\kappa) = \left| \Pr[\mathbf{Exp}_{CS,\mathcal{A}}^{\mathsf{SINV\text{-}CMA}}(1^\kappa) = 1] - \frac{1}{2} \right|,$$

where the probability is taken over all the random coins.

CS is $(t, \epsilon, q_s, q_v, q_{sc})$-SINV-CMA secure if there is no adversary operating in time t, issuing q_s queries to the signing oracle (followed potentially by queries to the `sconfirm` oracle), q_v queries to the confirmation/denial oracles and q_{sc}

queries to the selective conversion oracle that wins the game defined in Experiment $\mathbf{Exp}^{\mathsf{SINV\text{-}CMA}}(1^\kappa)$ with advantage greater that ϵ.

Theorem 4.2 (Invisibility of the New StE) *Given* $(t, q_s, q_v, q_{sc}) \in \mathbb{N}^4$ *and* $(\epsilon, \epsilon', \epsilon'') \in [0,1]^3$, *the new StE is* $(t, \epsilon, q_s, q_v, q_{sc})$*-*SINV-CMA *secure if it uses a* (t, ϵ', q_s)*-*SEUF-CMA *secure digital signature, a* $(t, \epsilon")$*-*INV-OT *secure DEM with injective encryption, and a* $(t + q_s(q_v + q_{sc}), \epsilon \cdot (1-\epsilon") \cdot (1-\epsilon')^{q_v+q_{sc}})$*-*IND-CPA *secure KEM.*

Proof Let $\mathcal{A}$ be an attacker that $(t, \epsilon, q_s, q_v, q_{sc})$-SINV-CMA breaks our construction. We construct an algorithm $\mathcal{R}$ that breaks the underlying KEM as follows.

[`keygen`] $\mathcal{R}$ gets the parameters of the KEM $\mathcal{K}$ from his challenger. Then, he chooses an appropriate $(t, \epsilon")$-INV-OT secure DEM $\mathcal{D}$ together with a (t, ϵ', q_s)-SEUF-CMA secure signature scheme Σ. $\mathcal{R}$ further generates a key pair $(\Sigma.sk, \Sigma.pk)$ for Σ and sets it as the signer's key pair.

[`sign` **and** `sconfirm` **queries**] For a signature query, $\mathcal{R}$ proceeds like the standard algorithm. He further maintains a list $\mathcal{L}$ of the encapsulations and corresponding keys used to generate the confirmer signatures.

[{`confirm`, `deny`} **queries**] For a verification query on a signature $\mu = (c, e, r)$ and a message m, $\mathcal{R}$ looks up the list $\mathcal{L}$ for the decapsulation of c, which once found, allows $\mathcal{R}$ to check the validity of the signature and therefore simulate correctly the suitable protocol (confirmation or denial). If c has not been used to generate confirmer signatures, then $\mathcal{R}$ will run the denial protocol. Note that $\mathcal{R}$ is able to perfectly simulate to confirmation/denial protocol since these are by definition concurrent zero-knowledge proofs.

This simulation differs from the real one when the signature $\mu = (c, e, r)$ on m is valid, but c does not appear in any record of $\mathcal{L}$. We distinguish two cases: either m was never queried to the signing oracle, then (m, μ) would correspond to an existential forgery on the confirmer signature (and thus to an existential forgery on Σ), or m has been previously queried to the signing oracle in which case (m, μ) would correspond to an existential forgery on Σ thanks to Remark 4.1. Hence, the probability that both scenarios do not happen is at least $(1-\epsilon')^{q_v}$ (Σ is (t, ϵ', q_s)-SEUF-CMA secure).

[`convert` **queries**] Conversion queries are treated like verification queries with the exception of converting the signature instead of running `confirm`, and issuing $\perp$ instead of running `deny`. Similarly, this simulation does not differ from the real execution of the algorithm with probability at least $(1-\epsilon')^{q_{sc}}$.

[Challenge] Eventually, $\mathcal{A}$ outputs a challenge message $m^\star$. $\mathcal{R}$ uses his challenge $(c^\star, k^\star)$ to compute a digital signature $(s^\star, r^\star)$ on $c^\star \| m^\star$. Then, he encrypts $s^\star$ in $e^\star$ using $\mathcal{D}.\mathtt{encrypt}_{k^\star}$ and outputs $\mu^\star = (c^\star, e^\star, r^\star)$ to $\mathcal{A}$. Therefore, $\mu^\star$ is either a valid confirmer signature on $m^\star$ or an element indistinguishable from a random element in the confirmer signature space ($k^\star$ is random and $\mathcal{D}$ is INV-OT secure, moreover $r^\star$ is information-theoretically independent from m and $\Sigma.pk$). If $\mu^\star$, in the latter case, is a random element in the confirmer signature space, then this complies with the scenario of a real attack. Otherwise, if $\mu^\star$ is *only* *indistinguishable from random*, then if the advantage of $\mathcal{A}$ is non-negligibly different from the advantage of an invisibility adversary in a real attack, then $\mathcal{A}$ can be easily used to (t, ϵ'')-INV-OT break $\mathcal{D}$. To sum up, under the INV-OT assumption of $\mathcal{D}$, that is with probability at least $1 - \epsilon$", the challenge confirmer signature $\mu^\star$ is either a valid confirmer signature on $m^\star$ or a random element in the confirmer signature space.

[Post challenge phase] $\mathcal{A}$ will continue issuing queries to the previous oracles, and $\mathcal{R}$ can answer as previously. Note that in this phase, $\mathcal{A}$ might request the verification or conversion of a confirmer signature $(c^\star, -, -)$ on a message $m_i \neq m^\star$. According to the previous analysis, such a signature is invalid w.r.t. m_i with probability at least $(1 - \epsilon')^{q_{sc}+q_v}$.

In case the verification/conversion query involves $m^\star$ and $c^\star$, then let $(c^\star, \tilde{e}, \tilde{r}) \neq \mu^\star$ be the queried signature. We have $(\tilde{e}, \tilde{r}) \neq (e^\star, r^\star)$. Two cases manifest. Either $r^\star = \tilde{r}$, in which case $(c^\star, \tilde{e}, \tilde{r})$ is invalid w.r.t. $m^\star$ since otherwise $e^\star$ and $\tilde{e}$ will be two different ciphertexts that decrypt to $s^\star$, which is impossible since $\mathcal{D}$ is by assumption a DEM with injective encryption. Or $r^\star \neq \tilde{r}$; therefore for $(c^\star, \tilde{e}, \tilde{r})$ to be valid w.r.t. $m^\star$, $(\mathcal{D}.\texttt{decrypt}(\tilde{e}), \tilde{r})$ $(\neq (s^\star, r^\star))$ must be a valid digital signature on $c^\star \| m^\star$. Therefore, this latter scenario does not happen with probability at least $(1-\epsilon')$ since Σ is SEUF-CMA secure.

Bottom line is, whenever a verification/conversion query involves $c^\star$ or an encapsulation c that is not in the list $\mathcal{L}$, $\mathcal{R}$ will issue the denial protocol in case of a verification query, or the symbol $\perp$ in case of a conversion query. The probability that the simulation does not differ from the real execution is at least $(1-\epsilon')^{q_{sc}+q_v}$.

[Final output] When $\mathcal{A}$ outputs his answer $b \in \{0, 1\}$, $\mathcal{R}$ will forward this answer to his own challenger. Therefore $\mathcal{R}$ will $(t+q_s(q_v+q_{sc}), \epsilon\cdot(1-\epsilon")\cdot(1-\epsilon')^{q_v+q_{sc}})$-IND-CPA break $\mathcal{K}$.

□

4.2 Practical Realizations

In this section, we provide practical realizations of confirmer signatures from the new StE. We first introduce some useful classes of the base primitives, then, we proceed to the description of concrete instantiations of the paradigm. It is worth noting that although we implement the verification protocols using zero-knowledge proofs of knowledge ZKPoK, zero-knowledge proofs work also in our setting. Actually, we resort to ZKPoK only to spare the effort of specifying the assumptions on the building blocks (signature and encryption) that ensure soundness; the recovery property of proofs of knowledge is not needed.

4.2.1 *The Class $\mathbb{S}$ of Signatures*

Definition 4.2 (Class $\mathbb{S}$ of Signatures) $\mathbb{S}$ is the set of all digital signatures for which there exists a pair of efficient algorithms, `convert` and `retrieve`, where `convert` inputs a public key pk, a message m, and a valid signature σ on m (according to pk) and outputs the pair (s, r) such that:

1. r reveals no information about m nor about pk, i.e. there exists an algorithm simulate such that for every public key pk from the key space and for every message m from the message space, the output simulate(pk, m) is statistically indistinguishable from r.
2. there exists an algorithm `compute` that on the input pk, the message m and r, computes a description of a function $f : (\mathbb{G}, *) \to (\mathbb{H}, \circ_s)$:
   - where $(\mathbb{G}, *)$ is a group and $\mathbb{H}$ is a set equipped with the binary operation $\circ_s$,
   - $\forall S, S' \in \mathbb{G}: f(S * S') = f(S) \circ_s f(S')$.

   and an $I \in \mathbb{H}$, such that $f(s) = I$.

and an algorithm `retrieve` that inputs pk, m and the correctly converted pair (s, r) and retrieves[1] the signature σ on m.

The class $\mathbb{S}$ differs from the class $\mathbb{C}$, introduced in Shahandashti and Safavi-Naini (2008), in the condition required for the function f. In fact, in our description of $\mathbb{S}$, the function f should satisfy a homomorphic property, whereas in the class $\mathbb{C}$, f should only possess an efficient zero-knowledge protocol for proving knowledge of preimages of values in its range. We show in Theorem 4.3 that signatures in $\mathbb{S}$ accept also efficient ZK proofs for proving knowledge of preimages, and thus belong to the class $\mathbb{C}$. Conversely, one can claim that signatures in $\mathbb{C}$ are also in $\mathbb{S}$, at least from a practical point of view, since it is not known in general how to achieve efficient ZK protocols for proving knowledge of preimages of f without having the latter item satisfy some homomorphic properties. It is worth noting that the class of signatures introduced in Goldwasser and Waisbard (2004) is similar to the classes $\mathbb{S}$ and $\mathbb{C}$; the difference is that the condition of having an efficient ZK protocol for proving knowledge of preimages is weakened to having only a WHPoK (recall that the prover in a WHPoK (Goldreich 2001, Sect. 4.6) does not reveal the witness but may leak some knowledge during his interaction with the verifier). Again, although this is a weaker assumption on f, all illustrations of signatures in this wider class happen to be also in $\mathbb{C}$ and $\mathbb{S}$. Our resort to specify the homomorphic property on f will be justified later when describing the confirmation/denial protocols of the resulting construction. In fact, these protocols are concurrent composition of proofs and therefore need a careful study as it is known that zero-knowledge is not closed under concurrent composition. Besides, the class $\mathbb{S}$ encompasses most proposals that

[1]Note that the `retrieve` algorithm suffices to ensure the non-triviality of the map f; given a pair (s, r) satisfying the conditions described in the definition, one can efficiently recover the original signature on the message.

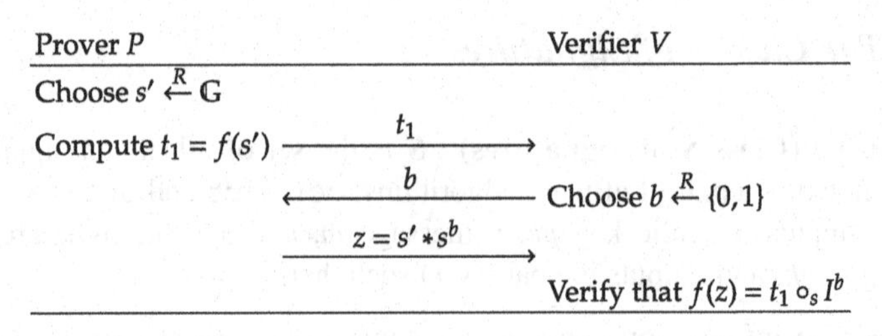
Prover P
Verifier V
Choose $s' \xleftarrow{R} \mathbb{G}$
Compute $t_1 = f(s')$
t_1
b
Choose $b \xleftarrow{R} \{0,1\}$
$z = s' * s^b$
Verify that $f(z) = t_1 \circ_s I^b$

Fig. 4.2 Proof of membership to the language $\{I{:}f(s) = I\}$ **Common input:** I and **Private input:** s

were suggested so far, e.g. Bellare and Rogaway (1996), Schnorr (1991), Gennaro et al. (1999), Boneh et al. (2004b), Pointcheval and Stern (2000), Cramer and Shoup (2000), Camenisch and Lysyanskaya (2002, 2004), Boneh and Boyen (2004), Zhang et al. (2004), and Waters (2005). The reason why $\mathbb{S}$ includes most digital signature schemes lies in the fact that a signature verification consists in applying a function f to the "vital" part of the signature in question, then comparing the result to an expression computed from the message underlying the signature, the "auxiliary" or "simulatable" part of the signature, and finally the public parameters of the signature scheme. The function f need not be one-way, however the signature scheme would be trivially forgeable if it is not the case. Moreover, f consists most of the time of an arithmetic operation (e.g. exponentiation, raising to a power, pairing computation) which easily satisfies a homomorphic property.

Theorem 4.3 *The protocol depicted in Fig. 4.2 is an efficient zero-knowledge proof of knowledge of preimages of the function f described in Definition 4.2.*

The proof is straightforward using the standard techniques. □

Remark 4.2 Soundness of the interactive proof in Fig. 4.2 is easily met when the map f is one-way.

4.2.2 The Class $\mathbb{E}$ of Encryption Schemes

Definition 4.3 (Class $\mathbb{E}$ of Encryption Schemes) $\mathbb{E}$ is the set of public-key encryption schemes Γ that have the following properties:

1. The message space is a group $\mathcal{M} = (\mathbb{G}, *)$ and the ciphertext space C is a set equipped with a binary operation $\circ_e$.
2. Let $m \in \mathcal{M}$ be a message and c its encryption with respect to a key pk. On the common input m, c, and pk, there exists an efficient zero-knowledge proof ZKP of m being the decryption of c with respect to pk. The private input of the prover is either the private key sk corresponding to pk, or the randomness used to encrypt m in c.

3. $\forall pk,\ \forall m, m' \in \mathcal{M}$:

$$\Gamma.\texttt{encrypt}_{pk}(m * m') = \Gamma.\texttt{encrypt}_{pk}(m) \circ_e \Gamma.\texttt{encrypt}_{pk}(m')$$

Moreover, given the randomnesses used to encrypt m in $\Gamma.\texttt{encrypt}_{pk}(m)$ and m' in $\Gamma.\texttt{encrypt}_{pk}(m')$, one can deduce (using only the public parameters) the randomness used to encrypt $m * m'$ in $\Gamma.\texttt{encrypt}_{pk}(m) \circ_e \Gamma.\texttt{encrypt}_{pk}(m')$.

Examples of encryption schemes in the above class include ElGamal's encryption (El Gamal 1985), Paillier's encryption (Paillier 1999), or Boneh-Boyen-Shacham's scheme (Boneh et al. 2004a). In fact, these schemes are homomorphic and possess an efficient proof of correctness of a decryption, namely the proof of equality of two discrete logarithms in case of El Gamal (1985) and Boneh et al. (2004a) and the proof of knowledge of an Nth root in case of Paillier (1999). Note that both ElGamal's and Boneh-Boyen-Shacham's encryptions are derived from the KEM/DEM paradigm and are therefore suitable for use in the new StE paradigm.

Theorem 4.4 *Let Γ be an encryption scheme from the above class $\mathbb{E}$. Let further e be an encryption, by Γ, of some message s under some public key pk. The protocol depicted in Fig. 4.3 is a zero-knowledge proof of knowledge of the decryption of e.*

Proof Completeness is straightforward.

Validity (knowledge extractability) is also easy. In fact, suppose a malicious prover $\tilde{P}$ can successfully answer two different challenges 0 and 1 (challenge space is $\{0, 1\}$) for the same commitment value t_2:

$$z_1 = \Gamma.\texttt{decrypt}(t_2) \wedge z_2 = \Gamma.\texttt{decrypt}(t_2 \circ_e e)$$

Since $\circ_e$ induces a group law in the ciphertext space of Γ, we have: $z_1^{-1} = \Gamma.\texttt{decrypt}(t_2^{-1})$. It follows that $\tilde{P}$ can compute a decryption of e as $z_1^{-1} * z_2 = \Gamma.\texttt{decrypt}(e)$. We conclude that the soundness error probability of the protocol is at most $1/2$ (we assume that ZKP has negligible soundness error). We will see

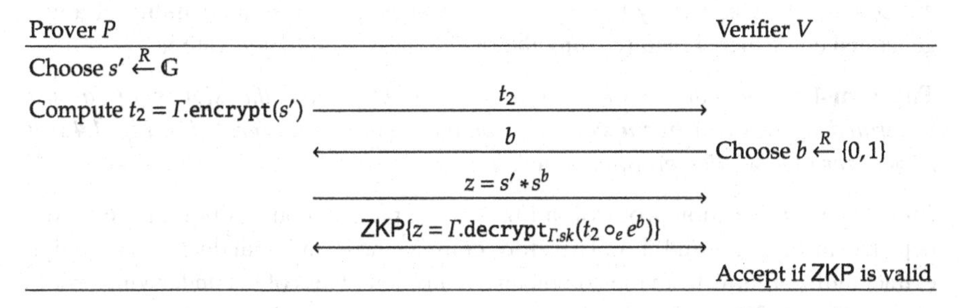

Prover P
Verifier V
Choose $s' \xleftarrow{R} G$
Compute $t_2 = \Gamma.\mathbf{encrypt}(s')$
t_2
b
Choose $b \xleftarrow{R} \{0,1\}$
$z = s' * s^b$
ZKP$\{z = \Gamma.\mathbf{decrypt}_{\Gamma.sk}(t_2 \circ_e e^b)\}$
Accept if ZKP is valid

Fig. 4.3 Proof of membership to the language $\{e\colon s = \Gamma.\texttt{decrypt}_{\Gamma.sk}(e)\}$ **Common input:** $(e, \Gamma.pk)$ and **Private input:** s and $\Gamma.sk$ or randomness encrypting s in e

in Sect. 4.3.1 how to reduce the soundness error without necessarily repeating the protocol many times.

For the zero-knowledgeness, we describe the following simulator:

1. Generate uniformly a random challenge $b' \stackrel{R}{\leftarrow} \{0,1\}$. Choose a random $z \stackrel{R}{\leftarrow} \mathbb{G}$, compute $t_2 = \Gamma.\mathtt{encrypt}_{\Gamma.pk}(z) \circ_e e^{-b'}$ and sends it to the verifier.
2. Get b from the verifier.
3. If $b = b'$, the simulator sends back z and simulates the proof ZKP for z being the decryption of $t_2 \circ_e e^b$ (this proof is simulatable since it is zero-knowledge by assumption). Otherwise, it goes to Step 2 (*rewinds* the verifier).

The prover's first message is always an encryption of a random value, and so is the first message of the simulator. Since b' is chosen uniformly at random from $\{0,1\}$, then, the probability that the simulator does not rewind the verifier is $1/2$, and thus the simulator runs in polynomial time in the security parameter. Finally, the distribution of the answers (last messages) of the prover and of the simulator is the same. We conclude that the proof is perfectly zero-knowledge. □

Remark 4.3 Soundness of the interactive proof depicted in Fig. 4.3 is ensured if the used encryption scheme is OW-CPA.

4.2.3 Confirmation/Denial Protocols

We combine a secure signature scheme $\Sigma \in \mathbb{S}$ and a secure encryption scheme $\Gamma \in \mathbb{E}$, which is *derived from the KEM/DEM paradigm*, in the way described in Sect. 4.1.1. Namely we first compute an encapsulation c together with its corresponding key k. Then we compute a signature σ on c concatenated with the message to be signed. Finally convert σ to (s, r) using the `convert` algorithm described in Definition 4.2 and encrypt s in $e = \mathcal{D}.\mathtt{encrypt}_k(s)$ using k. The resulting confirmer signature is (c, e, r). We describe in Fig. 4.4 the confirmation/denial protocols corresponding to the resulting construction. Note that the confirmation protocol can be also run by the signer who wishes to confirm the validity of a just generated confirmer signature using the randomness used to generate it.

Theorem 4.5 *The confirmation protocol (run by either the signer on a just generated signature or by the confirmer on any signature) described in Fig. 4.4 is a perfect zero-knowledge proof of knowledge.*

Proof The confirmation protocol in Fig. 4.4 is a parallel composition of the proofs depicted in Figs. 4.2 and 4.3. Therefore completeness and soundness (knowledge extractability) follow from the completeness and soundness of the underlying proofs (see Goldreich 2001). Finally, the ZK simulator is the parallel composition of the ZK simulators of the mentioned protocols. □

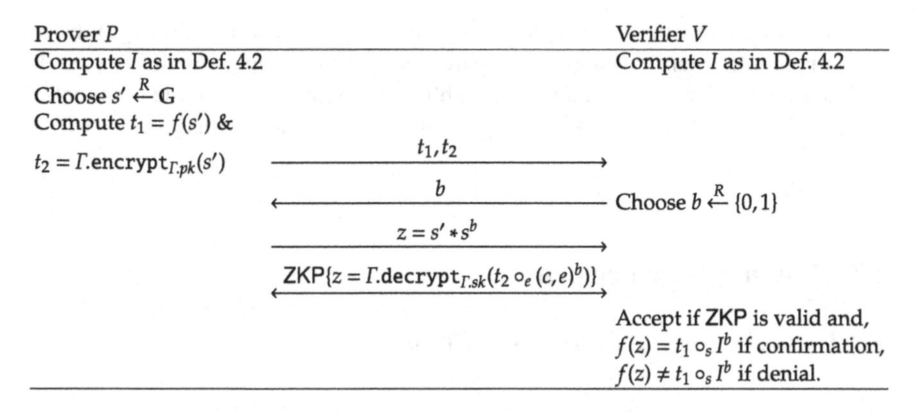

Prover P
Verifier V
Compute I as in Def. 4.2
Compute I as in Def. 4.2
Choose $s' \xleftarrow{R} \mathbb{G}$
Compute $t_1 = f(s')$ &
$t_2 = \Gamma.\mathtt{encrypt}_{\Gamma.pk}(s')$
t_1, t_2
b
Choose $b \xleftarrow{R} \{0,1\}$
$z = s' * s^b$
$\mathsf{ZKP}\{z = \Gamma.\mathtt{decrypt}_{\Gamma.sk}(t_2 \circ_e (c,e)^b)\}$
Accept if ZKP is valid and,
$f(z) = t_1 \circ_s I^b$ if confirmation,
$f(z) \neq t_1 \circ_s I^b$ if denial.

Fig. 4.4 Confirmation/denial protocol for the new StE: $\mathsf{PoK}\{s: s = \Gamma.\mathtt{decrypt}(c,e) \wedge \Sigma.\mathtt{verify}(\mathtt{retrieve}(s,r), m\|c) = (\neq)\ 1\}$ **Common input:** $(c, e, r, \Sigma.pk, \Gamma.pk)$ and **Private input:** $\Gamma.sk$ or randomness encrypting s in (c, e)

Theorem 4.6 *The denial protocol described in Fig. 4.4 is a proof of knowledge with computational zero-knowledge if Γ is* IND-CPA *secure.*

Proof Using the standard techniques, we prove that the denial protocol outlined in Fig. 4.4 is a complete and sound proof of knowledge. Similarly, we provide the following simulator to prove the ZK property.

1. Generate $b' \in_R \{0, 1\}$. Choose $z \in_R \mathbb{G}$ and a random $t_1 \in_R f(\mathbb{G})$ and $t_2 = \Gamma.\mathtt{encrypt}_{\Gamma.pk}(z) \circ_e (c, e)^{-b'}$.
2. Get b from the verifier. If $b = b'$, it sends z and simulates the proof ZKP of z being the decryption of $t_2 \circ_e (e, s_k)^b$. If $b \neq b'$, it goes to Step 1.

The prover's first message is an encryption of a random value $s' \in_R \mathbb{G}$, in addition to $f(s')$. The simulator's first message is an encryption of a random value $z * s^{-b'}$ and the element $t_1 \in_R f(\mathbb{G})$, which is *independent of* z. Distinguishing those two cases is at least as hard as breaking the IND-CPA security of the underlying encryption scheme. Therefore, under the IND-CPA security of the encryption scheme, the simulator's and prover's first message distributions are indistinguishable. Moreover, the simulator runs in expected polynomial time, since the number of rewinds is 2. Finally, the distributions of the prover's and the simulator's messages in the last round are again, by the same argument, indistinguishable under the IND-CPA security of the encryption scheme. □

Concurrent Zero-Knowledgeness If the proof ZKP underlying the above-mentioned protocols is a public-coin Honest-Verifier Zero-Knowledge (HVZK) protocol, then there are a number of efficient transformations that turn them into proofs with concurrent zero-knowledge, e.g. Micciancio and Petrank (2002). For instance, if ZKP is a Sigma protocol, then these confirmation/denial protocols can be efficiently turned into concurrent ZK proofs according to Damgård (2000); this transformation preserves the number of rounds while it incurs a tiny overhead in the computational complexity (computation of a commitment on a message). Note that although the transformation (Damgård 2000) is in the auxiliary string model, such a scenario is easy to achieve in a public-key setting; for example certificates computed by a PKI on public keys are possible candidates to auxiliary strings to the players.

4.3 Further Enhancements

4.3.1 Reducing the Soundness Error

The protocols presented earlier in the previous section consist of a proof of knowledge of preimages, by some *homomorphic* map, which incidentally satisfy a relation efficiently provable via a zero-knowledge proof ZKP.

In this section, we show how to reduce the soundness error of these protocols without necessarily repeating them. We will focus on the part of the protocol proving knowledge of the preimage; actually we assume ZKP has a negligible soundness error since it can itself implement the optimizations we propose if it is a proof of knowledge for group homomorphisms.

Let $f\colon (\mathbb{G}, *) \rightarrow (\mathbb{H}, \circ)$ be the homomorphic map underlying the proof of knowledge. Let further I be the value for which we want to prove knowledge of a preimage. We consider a challenge space C that satisfies, for some known values $\ell \in \mathbb{Z}$ and $u \in \mathbb{G}$, the following (Maurer 2015):

1. $\gcd(b_1 - b_2, \ell) = 1$ for all $b_1, b_2 \in C$ (with $b_1 \neq b_2$),
2. $f(u) = I^{\ell}$.

Note that the above conditions are easily met in groups with known prime order ℓ, i.e. discrete-log-based groups.

The protocol below is an efficient zero-knowledge proof of knowledge of a preimage of I, if C is polynomially bounded.

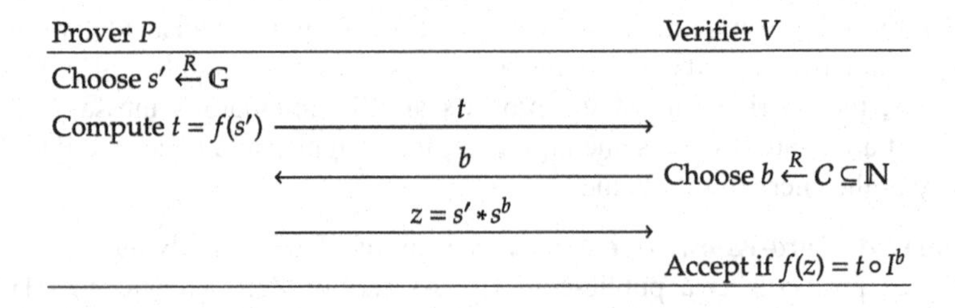

Completeness is straightforward.

For knowledge extractability, we consider two accepting transcripts for the same commitment value t and different challenges b_1, b_2 ($b_2 \geq b_1$). Let z_1, z_2 the respective responses of the prover in the last round.

We have $f(z_1) = t \circ I^{b_1}$ and $f(z_2) = t \circ I^{b_2}$. Therefore $f(z_1^{-1} * z_2) = I^{b_2 - b_1}$. We compute values x, y by the Extended Euclidean Algorithm to get. $x\ell + y(b_2 - b_1) = 1$. It follows that $I = I^{x\ell} I^{y(b_2 - b_1)}$. Thus $I = f(u^x * (z_1^{-1} * z_2)^y)$. In other words, a preimage of I can be computed as $u^x * (z_1^{-1} * z_2)^y$.

Finally, the ZK simulator is similar to that of the original protocol with the exception of drawing the challenge b', in the first stage of the protocol, from C. The new probability of not rewinding the verifier is $1/|C|$. Thus, C must be polynomially bounded in order to guarantee a polynomial running time of the simulator.

4.3.2 Online Non-transferability

Online non-transferability as previously mentioned allows to avoid some attacks in which the intended verifier, say V, interacts *concurrently* with the genuine prover and a hidden malicious verifier $\widetilde{V}$ such that this latter gets convinced of the proven statement (validity or invalidity of a signature w.r.t. a given message).

One way to circumvent this problem consists in using designated-verifier proofs (Jakobsson et al. 1996). In fact, these proofs can be conducted by both the prover and the verifier. When a verifier receives such a proof, he will be convinced of the validity of the underlying statement since he has not proved it himself. However, he cannot convince a third party of the validity of the statement as he can himself perfectly simulate the answers sent by the prover.

A generic construction of designated-verifier proofs from Σ protocols (Sect. 1.4.3) was given in Shahandashti and Safavi-Naini (2008). The idea consists in proving either the statement in question or proving knowledge of the verifier's private key. This is achieved using *proofs of disjunctive knowledge* if the proof of the statement and the proof of knowledge of the verifier's private key are both Σ protocols. More precisely, let the statement to be proven be of the form $(x, w) \in \mathcal{R}$ where $\mathcal{R}$ is some NP relation, and (x, w) is the pair (instance,witness). Let further (sk_V, pk_V) be the verifier's key pair. Suppose that both the statements $(x, w) \in \mathcal{R}$ and $(sk_V, pk_V) \in \mathsf{Valid}$ can be proven using Σ protocols. The work-flow of the proof:

$$\mathsf{PoK}\{(x, w) \in \mathcal{R} \ \vee \ (sk_V, pk_V) \in \mathsf{Valid}\}$$

is depicted in Fig. 4.5.

It is clear that this protocol can be also carried out using the private input sk_V. We refer to Shahandashti and Safavi-Naini (2008) for further details about the analysis of this protocol.

Getting back to our problem, the already mentioned confirmation/denial protocols can be shown to be Σ protocols if the proof ZKP in the last round is non-interactive (this would necessitate the presence of a trusted authority). In this case, they can be efficiently transformed into designated-verifier proofs, providing therefore the required online non-transferability for the resulting signatures. Note that if we consider such proofs in the registered key model, i.e. a model that requires

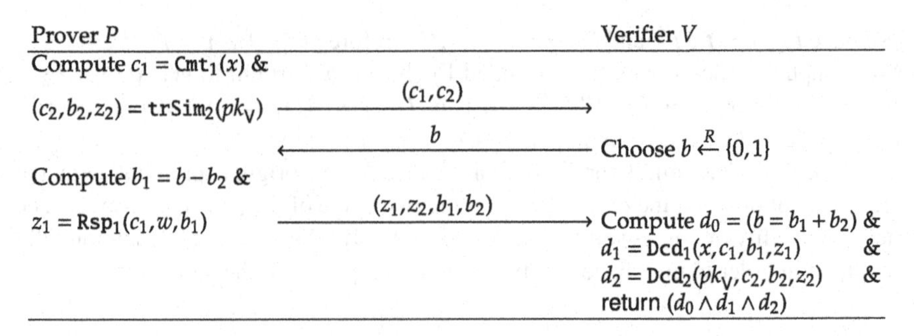

Fig. 4.5 Proof of disjunctive knowledge: $\mathsf{PoK}\{(x, w) \in \mathcal{R} \vee (sk_V, pk_V) \in \mathsf{Valid}\}$ **Common input**: (x, pk_V) and **Private input**: w or sk_V

the verifier to prove knowledge of his private key before a certifying authority, the reduction can easily run these protocols for the invisibility adversary given the knowledge of the verifier's private key.

4.4 Performance of the New StE

The variant of StE presented in this chapter improves the plain paradigm (Camenisch and Michels 2000) as it weakens the assumption on the underlying encryption from IND-PCA to IND-CPA. This impacts positively the efficiency of the construction from many sides. In fact, the resulting signature is shorter and its generation/verification cost is smaller. Moreover, the confirmation/denial protocols are rendered more efficient by allowing homomorphic encryption schemes as shown earlier in this section, e.g. El Gamal (1985), Boneh et al. (2004a).[2] Such encryption schemes were not possible to use before since a homomorphic scheme can never attain the IND-PCA security. Besides, even when the IND-PCA encryption scheme accepts efficient proofs of knowledge of decryptions, e.g. Abdalla et al. (2015), the involved protocols are more expensive than those corresponding to their IND-CPA variant.

The construction achieves also better performances than the proposal of Goldwasser and Waisbard (2004), where the confirmer signature comprises k commitments and $2k$ IND-CCA encryptions, where k is the number of rounds used in the confirmation protocol. Moreover, the denial protocol presented in Goldwasser and Waisbard (2004) suffers the resort to proofs of general NP statements (where the considered encryption is IND-CCA). The same remark applies to the construction of Wikström (2007) where both the confirmation and denial protocols rely on proofs of general NP statements.

[2]Both schemes are IND-CPA secure and are derived from the KEM/DEM paradigm. Moreover, the underlying KEM and DEM present interesting homomorphic properties that make them belong to the class $\mathbb{E}$ of encryption schemes. We refer to the discussion after Definition 4.3 for the details.

Finally, we remark that this variant of StE, first introduced in El Aimani (2008), captures many efficient realizations of confirmer/undeniable signatures, e.g. Le Trieu et al. (2010), Schuldt and Matsuura (2010). It also serves for analyzing some early schemes that had a speculative security: the Damgård-Pedersen undeniable signatures (Damgård and Pedersen 1996). In fact, we showed in Sect. 3.6, that these signatures are unlikely to be invisible, and we proposed a fix so that they meet the required security notion; interestingly, this repair turns out to be a special instantiation of the new StE paradigm.

References

Abdalla M, Benhamouda F, Pointcheval D (2015) Public-key encryption indistinguishable under plaintext-checkable attacks. In: Katz J (ed) PKC. LNCS, vol 9020. Springer, Heidelberg, pp 332–352

Bellare M, Rogaway P (1996) The exact security of digital signatures: how to sign with RSA and Rabin. In: Maurer UM (ed) (1996) Proceeding of advances in cryptology - EUROCRYPT'96, international conference on the theory and application of cryptographic techniques, Saragossa, 12–16 May 1996. LNCS, vol 1070. Springer, Heidelberg, pp 399–416

Boneh D, Boyen X (2004) Short signatures without random Oracles. In: Cachin C, Camenisch J (eds) Advances in cryptology - EUROCRYPT 2004. LNCS, vol 3027. Springer, Heidelberg, pp 56–73

Boneh D, Boyen X, Shacham H (2004a) Short group signatures. In: Franklin MK (ed) (2004) Proceedings of advances in cryptology - CRYPTO 2004, 24th annual international cryptology conference, Santa Barbara, CA, 15–19 August 2004. LNCS, vol 3152. Springer, Heidelberg, pp 41–55

Boneh D, Lynn B, Shacham H (2004b) Short signatures from the Weil pairing. J Cryptol 17(4):297–319

Camenisch J, Lysyanskaya A (2002) Dynamic accumulators and application to efficient revocation of anonymous credentials. In: Yung M (ed) CRYPTO. LNCS, vol 2442. Springer, Heidelberg, pp 61–76

Camenisch J, Lysyanskaya A (2004) Signature schemes and anonymous credentials from bilinear maps. In: Franklin MK (ed) (2004) Proceedings of advances in cryptology - CRYPTO 2004, 24th annual international cryptology conference, Santa Barbara, CA, 15–19 August 2004. LNCS, vol 3152. Springer, Heidelberg, pp 56–72

Camenisch J, Michels M (2000) Confirmer signature schemes secure against adaptative adversaries. In: Preneel B (ed) (2000) Proceeding of advances in cryptology - EUROCRYPT 2000, international conference on the theory and application of cryptographic techniques, Bruges, 14–18 May 2000. LNCS, vol 1807. Springer, Heidelberg, pp 243–258

Cramer R, Shoup V (2000) Signature schemes based on the strong RSA assumption. ACM Trans Inf Syst Secur 3(3):161–185

Damgård IB (2000) Efficient concurrent zero-knowledge in the auxiliary string model. In: Preneel B (ed) (2000) Proceeding of advances in cryptology - EUROCRYPT 2000, international conference on the theory and application of cryptographic techniques, Bruges, 14–18 May 2000. LNCS, vol 1807. Springer, Heidelberg, pp 418–430

Damgård IB, Pedersen TP (1996) New convertible undeniable signature schemes. In: Maurer UM (ed) (1996) Proceeding of advances in cryptology - EUROCRYPT'96, international conference on the theory and application of cryptographic techniques, Saragossa, 12–16 May 1996. LNCS, vol 1070. Springer, Heidelberg, pp 372–386

El Aimani L (2008) Toward a generic construction of universally convertible undeniable signatures from pairing-based signatures. In: Chowdhury DR, Rijmen V, Das A (eds) Progress in cryptology - INDOCRYPT 2008. LNCS, vol 5365. Springer, Heidelberg, pp 145–157. Full version available at the Cryptology ePrint Archive, Report 2009/362

El Gamal T (1985) A public key cryptosystem and a signature scheme based on discrete logarithms. IEEE Trans Inf Theory 31:469–472

Galbraith SD, Mao W (2003) Invisibility and anonymity of undeniable and confirmer signatures. In: Joye M (ed) Topics in cryptology - CT-RSA 2003. LNCS, vol 2612. Springer, Heidelberg, pp 80–97

Gennaro R, Halevi S, Rabin T (1999) Secure hash-and-sign signatures without the random Oracle. In: Stern J (ed) (1999) Proceeding of advances in cryptology - EUROCRYPT'99, international conference on the theory and application of cryptographic techniques, Prague, 2–6 May 1999. LNCS, vol 1592. Springer, Heidelberg, pp 397–416

Goldreich O (2001) Foundations of cryptography. Basic tools. Cambridge University Press, Cambridge

Goldwasser S, Waisbard E (2004) Transformation of digital signature schemes into designated confirmer signature schemes. In: Naor M (ed) Theory of cryptography, TCC 2004. LNCS, vol 2951. Springer, Heidelberg, pp 77–100

Jakobsson M, Sako K, Impagliazzo R (1996) Designated verifier proofs and their applications. In: Maurer UM (ed) (1996) Proceeding of advances in cryptology - EUROCRYPT'96, international conference on the theory and application of cryptographic techniques, Saragossa, 12–16 May 1996. LNCS, vol 1070. Springer, Heidelberg, pp 143–154

Le Trieu P, Kurosawa K, Ogata W (2010) Provably secure convertible undeniable signatures with unambiguity. In: Garay JA, Prisco RD (eds) SCN 2010. LNCS, vol 6480. Springer, Heidelberg. Full version available at the Cryptology ePrint Archive, Report 2009/394

Maurer UM (2015) Zero-knowledge proofs of knowledge for group homomorphisms. Des Codes Cryptogr 77(2–3):663–676

Micciancio D, Petrank E (2002) Efficient and concurrent zero-knowledge from any public coin HVZK protocol. In: Electronic Colloquium on Computational Complexity (ECCC) (045)

Paillier P (1999) Public-key cryptosystems based on composite degree residuosity classes. In: Stern J (ed) (1999) Proceeding of advances in cryptology - EUROCRYPT'99, international conference on the theory and application of cryptographic techniques, Prague, 2–6 May 1999. LNCS, vol 1592. Springer, Heidelberg, pp 223–238

Pointcheval D, Stern J (2000) Security arguments for digital signatures and blind signatures. J Cryptol 13(3):361–396

Schnorr CP (1991) Efficient signature generation by smart cards. J Cryptol 4(3):161–174

Schuldt JCN, Matsuura K (2010) An efficient convertible undeniable signature scheme with delegatable verification. In: Kwak J, Deng RH, Won Y, Wang G (eds) ISPEC 2010. LNCS, vol 6047. Springer, Heidelberg, pp 276–293. Full version available at the Cryptology ePrint Archive, Report 2009/454

Shahandashti SF, Safavi-Naini R (2008) Construction of universal designated-verifier signatures and identity-based signatures from standard signatures. In: Cramer R (ed) PKC 2008. LNCS, vol 4939. Springer, Heidelberg, pp 121–140

Waters B (2005) Efficient identity-based encryption without random Oracles. In: Cramer R (ed) Advances in cryptology - EUROCRYPT 2005. LNCS, vol 3494. Springer, Heidelberg, pp 114–127

Wikström D (2007) Designated confirmer signatures revisited. In: Vadhan SP (ed) TCC 2007. LNCS, vol 4392. Springer, Heidelberg, pp 342–361

Zhang F, Safavi-Naini R, Susilo W (2004) An efficient signature scheme from bilinear pairings and its applications. In: Bao F, Deng RH, Zhou J (eds) 7th international workshop on practice and theory in public key cryptography, PKC 2004. LNCS, vol 2947. Springer, Heidelberg, pp 277–290

Part III
The “Commit_then_Encrypt_and_Sign” (CtEaS) Paradigm

Chapter 5
Analysis of CtEaS

Abstract Efficient as the (new) StE is, it can only be used with a restricted class of signatures in order to allow effective verification. The Commit_then_Encrypt_and _Sign (CtEaS) paradigm has the merit of accepting *any* signature among its building blocks without compromising the verification protocols. In this chapter, we investigate this method by determining the exact security property needed for the encryption to achieve secure constructions. Our study, conducted for confirmer signatures, applies also to signcryption.

5.1 CtEaS for Confirmer Signatures

In addition to accepting any signature scheme in its base primitives, CtEaS has the advantage of performing signature and encryption *in parallel*, in contrast to the sequential composition of StE.

CtEaS includes among its building blocks a public-key encryption scheme that supports labels, which was used in Wang et al. (2007) in order to fix a flaw that afflicted the original proposal in Gentry et al. (2005). More precisely, a confirmer signature on a message m is obtained by first committing to m, then encrypting the randomness used in the commitment w.r.t. the label $m \| \Sigma.pk$ ($\Sigma.pk$ is the public key of the used digital signature scheme), *and* finally signing the commitment.

However, we remark that this repair will violate the invisibility of the resulting construction. In fact, the standard security definitions for encryption with labels do not require the label of a ciphertext to be hidden (since the label is required as input to the decryption algorithm in order to correctly decrypt the ciphertext). This implies that the signed message will be leaked from the encryption of the randomness used for the commitment. We can remediate this problem by using public-key encryption without labels to encrypt both the randomness and $m \| \Sigma.pk$. We describe in Fig. 5.1 our revised variant of this paradigm. The construction uses the following building blocks:

- **A digital signature scheme** Σ given by (1) $\Sigma.\texttt{keygen}$ which generates a key pair $(\Sigma.sk, \Sigma.pk)$ (2) $\Sigma.\texttt{sign}$ (3) $\Sigma.\texttt{verify}$.
- **A public-key encryption scheme** Γ described by (1) $\Gamma.\texttt{keygen}$ that generates the key pair $(\Gamma.sk, \Gamma.pk)$ (2) $\Gamma.\texttt{encrypt}$ (3) $\Gamma.\texttt{decrypt}$.

L. El Aimani, *Verifiable Composition of Signature and Encryption*,
https://doi.org/10.1007/978-3-319-68112-2_5

$\mathtt{setup}(1^\kappa)$	: $\Sigma.\mathtt{setup}(1^\kappa)$; $\Gamma.\mathtt{setup}(1^\kappa)$; $\Omega.\mathtt{setup}(1^\kappa)$
$\mathtt{keygen_S}(1^\kappa)$	: $\Sigma.\mathtt{keygen}(1^\kappa)$
$\mathtt{keygen_C}(1^\kappa)$	: $\Gamma.\mathtt{keygen}(1^\kappa)$
$\mathtt{sign}(m)$	: $c \leftarrow \Omega.\mathtt{commit}(m,r)$; $e \leftarrow \Gamma.\mathtt{encrypt}_{\{\Gamma.pk,coins_e\}}(r\|m\|\Sigma.pk)$
	$\sigma \leftarrow \Sigma.\mathtt{sign}_{\Sigma.sk}(c)$; return (c,e,σ)
$\mathtt{sconfirm}(\{c,e,\sigma\},m)$	: $\mathsf{ZKP}\Big\{(r,coins_e): c = \Omega.\mathtt{commit}(m,r) \wedge$
	$e = \Gamma.\mathtt{encrypt}_{\{\Gamma.pk,coins_e\}}(r\|m\|\Sigma.pk)\Big\}$
$\mathtt{confirm}(\{c,e,\sigma\},m)$	: $\mathsf{ZKP}\Big\{(r,\Gamma.sk): c = \Omega.\mathtt{commit}(m,r) \wedge r\|m\|\Sigma.pk = \Gamma.\mathtt{decrypt}_{\Gamma.sk}(e)\Big\}$
$\mathtt{deny}(\{c,e,\sigma\},m)$	: $\mathsf{ZKP}\Big\{(r,\Gamma.sk): c \neq \Omega.\mathtt{commit}(m,r) \wedge r\|m\|\Sigma.pk = \Gamma.\mathtt{decrypt}_{\Gamma.sk}(e)\Big\}$
$\mathtt{convert}(\{c,e,\sigma\},m)$	: $r\|m\|\Sigma.pk \leftarrow \Gamma.\mathtt{decrypt}_{\Gamma.sk}(c)$; $b \leftarrow (c = \Omega.\mathtt{commit}(m,r))$
	if $b = 0$ return $(\perp)$ else return (r,c,σ)

Fig. 5.1 The CtEaS paradigm

- **A commitment scheme** Ω given by (1) Ω.commit (2) Ω.open.

Note that sconfirm, confirm, and deny are only carried out when the signature σ on the commitment c is valid, otherwise the confirmer signature $\mu = (c, e, \sigma)$ is clearly deemed invalid w.r.t. m.

Remark 5.1 It is possible to require a proof, in the convert algorithm of CtEaS, that the revealed information is indeed a correct decryption of the corresponding encryption; such a proof is again possible to issue (with or without interaction) since the underlying statement is in NP.

Theorem 5.1 (Unforgeability of CtEaS) *Given* $(t, q_s) \in \mathbb{N}^2$ *and* $(\epsilon, \epsilon') \in [0, 1]^2$, *the construction depicted above is* $(t, \epsilon \cdot (1 - \epsilon')^{q_s}, q_s)$-EUF-CMA *secure if it uses a* (t, ϵ')*-binding commitment and a* (t, ϵ, q_s)-EUF-CMA *secure digital signature scheme.*

Proof Let $\mathcal{A}$ be an EUF-CMA attacker against the construction. We construct an EUF-CMA attacker $\mathcal{R}$ against the underlying digital signature scheme as follows.

$\mathcal{R}$ gets the parameters of the digital signature from her attacker, and chooses suitable encryption and commitment schemes. After she gets the confirmer's key pair from $\mathcal{A}$, $\mathcal{R}$ can perfectly simulate signature queries using her own challenger. At some point, $\mathcal{A}$ will output a forgery $\mu^\star = (c^\star, e^\star, \sigma^\star)$ on some message $m^\star$, which was never queried before for signature. By definition, $\sigma^\star$ is a valid digital signature on $c^\star$. Suppose there exists $1 \le i \le q_s$ such that $c^\star = c_i$ where $\mu_i = (c_i, e_i, \sigma_i)$ was the output confirmer signature on the query m_i. With probability at least $(1 - \epsilon')$, we have $m_i = m^\star$ (the commitment is (t, ϵ')-binding), which is a contradiction. Therefore, $c^\star$ was not queried by $\mathcal{R}$ for a digital signature with probability at least $(1 - \epsilon')^{q_s}$. $\mathcal{R}$ outputs to her challenger the EUF-CMA forgery $\sigma^\star$ and $c^\star$. □

5.2 The Exact Invisibility of CtEaS

We investigate in this section invisibility of confirmer signatures from CtEaS, i.e. we determine the exact security notion required for the underlying encryption to build invisible confirmer signatures.

Similarly to StE, we start by ruling out the notions OW-CPA, OW-PCA, and IND-CPA by remarking that ElGamal's encryption meets all those notions but cannot be used as it induces an invisibility infringement due to its homomorphic nature. Next, we exclude OW-CCA and NM-CPA using the meta-reduction tool. The subsequent notion to be considered is IND-PCA; luckily it turns out to be sufficient to achieve invisibility if combined with suitable signature and commitment schemes.

5.2.1 Impossibility Results

5.2.1.1 Deficiency of Homomorphic Encryption

Fact 5.1 *The CtEaS paradigm cannot lead to* INV-CMA *secure confirmer signatures when used with homomorphic encryption.*

Proof Let m_0, m_1 be the challenge messages the invisibility adversary $\mathcal{A}$ outputs to his challenger. Let further Γ, Σ, and Ω denote respectively the homomorphic encryption, the digital signature, and the commitment used as building blocks.

$\mathcal{A}$ gets as a challenge confirmer signature some $\mu_b = [c = \Omega.\texttt{commit}(m_b, r), e, \Sigma.\texttt{sign}(c)]$ ($b \in \{0, 1\}$) where e is an encryption of $r \| m_b \| \Sigma.pk$. $\mathcal{A}$ computes a new confirmer signature on m_b by multiplying e with an encryption of the identity element (of the message space of Γ), then queries this new signature (w.r.t. either m_0 or m_1) for conversion/verification; the response is sufficient for $\mathcal{A}$ to conclude. □

Corollary 5.1 *Invisibility in CtEaS cannot rest on* OW-CPA, OW-PCA, *or* IND-CPA *secure encryption.* □

5.2.1.2 Insufficiency of OW-CCA Secure Encryption

Lemma 5.1 *Assume there exists a key-preserving reduction $\mathcal{R}$ that converts an* INV-CMA *adversary $\mathcal{A}$ against confirmer signatures from CtEaS to a* OW-CCA *adversary against the underlying encryption scheme. Then, there exists a meta-reduction $\mathcal{M}$ that* OW-CCA *breaks the encryption scheme in question.*

Proof Let $\mathcal{R}$ be the key-preserving reduction that reduces OW-CCA breaking the encryption scheme underlying the construction to INV-CMA breaking the construction itself. We will construct an algorithm $\mathcal{M}$ that uses $\mathcal{R}$ to OW-CCA break the same encryption scheme by simulating an execution of the INV-CMA adversary $\mathcal{A}$ against the construction.

Let Γ be the encryption scheme $\mathcal{M}$ is trying to attack w.r.t. key $\Gamma.pk$. $\mathcal{M}$ proceeds as follows.

$\mathcal{M}$ launches $\mathcal{R}$ over Γ with the same key $\Gamma.pk$ and the same challenge e. Thus, all decryption queries made by $\mathcal{R}$, which are by definition different from the challenge e, can be forwarded to $\mathcal{M}$'s own challenger.

At some point, $\mathcal{M}$, acting as an INV-CMA attacker against the construction, outputs two challenge messages m_0, m_1 (chosen randomly from the message space) and gets as response a challenge $\mu_b = (c_b, e_b, \sigma_b)$ which is, with noticeable probability, a valid confirmer signature on m_b for some $b \in \{0, 1\}$. $\mathcal{M}$ is asked to find b.

We first note that $e_b \neq e$ holds with overwhelming probability; the challenge e does not encrypt a string whose suffix is $m_0 \| \Sigma.pk$ or $m_1 \| \Sigma.pk$ (although $\Sigma.pk$ can be maliciously chosen by $\mathcal{R}$, m_0 and m_1 are independently chosen by $\mathcal{M}$ upon receipt of the challenge e). Therefore, $\mathcal{M}$ requests his own challenger for the decryption of e_b. The answer to such a query will allow $\mathcal{M}$ (behaving as an INV-CMA attacker) to perfectly answer his invisibility challenge. $\mathcal{R}$ is then expected to return the answer to the OW-CCA challenge. Upon receipt of this answer, $\mathcal{M}$ will forward it to his own challenger. □

5.2.1.3 Insufficiency of NM-CPA Secure Encryption

Lemma 5.2 *Assume there exists a key-preserving reduction $\mathcal{R}$ that converts an* INV-CMA *adversary $\mathcal{A}$ against confirmer signatures from CtEaS to an* NM-CPA *adversary against the underlying encryption scheme. Then, there exists a meta-reduction $\mathcal{M}$ that* NM-CPA *breaks the encryption scheme in question.*

Proof Let $\mathcal{R}$ be the key-preserving reduction that reduces NM-CPA breaking the encryption underlying the construction to INV-CMA breaking the CtEaS construction. We construct an algorithm $\mathcal{M}$ that uses $\mathcal{R}$ to NM-CPA break the same encryption scheme by simulating an execution of the INV-CMA adversary $\mathcal{A}$ against the construction.

Let Γ be the encryption scheme $\mathcal{M}$ is trying to attack w.r.t. public key $\Gamma.pk$. $\mathcal{M}$ will launch $\mathcal{R}$ over the same public key $\Gamma.pk$. Next, $\mathcal{M}$ will simulate an INV-CMA adversary against the construction as follows.

$\mathcal{M}$ queries $\mathcal{R}$ on m_0, m_1 ($m_0 \neq m_1$) for confirmer signatures. Let $\mu_0 = (c_0, e_0, \sigma_0)$ and $\mu_1 = (c_1, e_1, \sigma_1)$ be the corresponding confirmer signatures. $\mathcal{M}$ queries again μ_0, μ_1, along with the corresponding messages, for conversion. Let r_0 and r_1 be the randomnesses used to generate the commitments c_0 and c_1 on m_0 and m_1 resp. $\mathcal{M}$ inputs $\mathcal{D} = \{r_0 \| m_0 \| \Sigma.pk, r_1 \| m_1 \| \Sigma.pk\}$ to his own challenger as a distribution probability from which the plaintexts will be drawn. $\mathcal{M}$ will receive as a challenge encryption $e^\star$. At that point, $\mathcal{M}$ chooses a bit $b \xleftarrow{R} \{0, 1\}$, and queries $\mathcal{R}$ on $\mu^\star = (c_b, e^\star, \sigma_b)$ and the message m_b for conversion. Note that if $e^\star$ encrypts

$r_b \| m_b \| \Sigma.pk$, then $\mu^\star$ is a valid confirmer signature on m_b, otherwise it is invalid (on either m_0 or m_1). Therefore, if the outcome of the query is not $\perp$, then $\mathcal{M}$ outputs $\Gamma.\texttt{encrypt}_{pk}(\overline{r_b})$, where $\overline{r_b}$ refers to the bit-complement of r_b, and the relation R: $R(r, r') = (r' = \overline{r})$. Otherwise, $\mathcal{M}$ outputs $\Gamma.\texttt{encrypt}_{pk}(\overline{r_{1-b}})$ and the same relation R. Finally $\mathcal{M}$ aborts the INV-CMA game. □

Theorem 5.2 *Consider the security notions obtained from pairing a security goal* GOAL $\in$ {OW, IND, NM} *and an attack model* ATK $\in$ {CPA, PCA, CCA}.
To achieve invisibility in CtEaS, the underlying encryption must be, in case the considered reduction is key-preserving, at least IND-PCA *secure. The restriction on the reduction can be lifted if the encryption enjoys non-malleability of the key generator.* □

5.2.2 *Sufficiency of* IND-PCA *Secure Encryption*

Likewise, we explain the above impossibility result by the *strong forgeability* of CtEaS. In fact, an adversary $\mathcal{A}$, given a valid signature $\mu = (\mu_1, \mu_2, \mu_3)$ on a message m, can create another valid signature μ' on m without the help of the signer as follows: $\mathcal{A}$ first requests the selective conversion of μ to obtain the decryption of μ_2, say $r\|m\|\Sigma.pk$, which he re-encrypts in μ_2'. Obviously, $\mu' = (\mu_1, \mu_2', \mu_3)$ is also a valid confirmer signature on m that the signer did not produce, and thus cannot confirm/deny or convert without having access to a decryption oracle of the encryption scheme underlying the construction. This explains the insufficiency of notions like IND-CPA. However, we observe that IND-CCA secure encryption is more than needed in this framework since a query of the type μ' is not completely uncontrolled by the signer. In fact, its second component μ_2' is an encryption of some data already disclosed by the signer, and thus a plaintext-checking oracle is sufficient to deal with such a query if the used digital signature is SEUF-CMA secure.

Theorem 5.3 (Invisibility of CtEaS) *Given* $(t, q_s, q_v, q_{sc}) \in \mathbb{N}^4$ *and* $(\epsilon, \epsilon', \epsilon_b) \in [0, 1]^3$, *confirmer signatures from CtEaS are* $(t, \epsilon, q_s, q_v, q_{sc})$-INV-CMA *secure if they use a* (t, ϵ', q_s)-SEUF-CMA *secure digital signature, a statistically hiding and* (t, ϵ_b) *binding commitment, and a*
$(t + q_s q_{sc}(q_{sc} + q_v), \frac{\epsilon}{2} \cdot [(1-\epsilon') \cdot (1-\epsilon_b)]^{(q_{sc}+q_v)}, q_{sc}(q_{sc} + q_v))$-IND-PCA *secure encryption scheme.*

Proof Let $\mathcal{A}$ be an attacker against the CtEaS construction. We construct an attacker $\mathcal{R}$ against the underlying encryption:

[`setup` **and** `keygen`] $\mathcal{R}$ gets the parameters of the encryption scheme Γ from her challenger. Then she chooses a (t, ϵ', q_s)-SEUF-CMA digital signature Σ (along with a key pair $(\Sigma.pk, \Sigma.sk)$) and a secure commitment Ω.

[`sign` **and** `sconfirm` **queries**] $\mathcal{R}$ proceeds exactly like the standard algorithm/protocol, with the exception of maintaining, in case of `sign`, in a list $\mathcal{L}$ the queried messages, the corresponding confirmer signatures, and the intermediate values used to produce these, for instance the random strings used to produce the commitments.

[`convert` **and** {`confirm,deny`} **queries**] To convert an alleged signature $\mu_i = (c_i, e_i, \sigma_i)$ on a message m_i, $\mathcal{R}$ checks the validity of σ_i on c_i; if it is invalid, then $\mathcal{R}$ proceeds as prescribed by the standard algorithm. Otherwise, $\mathcal{R}$ checks the list $\mathcal{L}$ for records corresponding to the queried message m_i and where c_i has been used as a commitment on m_i. If e_i is found in one of these records as encryption of some r_i concatenated with $m_i \| \Sigma.pk$ (r_i is the opening value of c_i), then $\mathcal{R}$ proceeds as dictated by the standard algorithm. Otherwise, $\mathcal{R}$ queries her PCA oracle on e_i and on each opening value of c_i found in these records (concatenated always with $m_i \| \Sigma.pk$). $\mathcal{R}$ returns the opening value giving rise to a 'yes' response (by the PCA oracle), if any, otherwise she returns $\perp$.

Verification ({`confirm`, `deny`}) queries are handled similarly with the exception of simulating the denial protocol instead of returning $\perp$, and the confirmation protocol instead of converting the signature.

This simulation differs from the standard procedure when μ_i is valid, but m_i has not been queried before, or c_i has not been used to generate commitments on m_i. The first case corresponds to an existential forgery on the construction which translates into breaking the binding property of the commitment scheme if c_i has been used as a commitment on some message $m_j \neq m_i$, or to breaking the existential unforgeability of the underlying digital signature otherwise. The second case corresponds to an existential forgery on the underlying signature scheme. Both cases do not happen with probability at least $[(1-\epsilon') \cdot (1-\epsilon_b)]^{q_v+q_{sc}}$.

[Challenge phase] At some point, $\mathcal{A}$ outputs two messages m_0, m_1 to $\mathcal{R}$. The latter chooses a random string r from the corresponding space. $\mathcal{R}$ outputs to her challenger the strings $r \| m_0 \| \Sigma.pk$ and $r \| m_1 \| \Sigma.pk$. She receives then a ciphertext e_b, encryption of $r \| m_b \| \Sigma.pk$, for some $b \in \{0, 1\}$. To answer her challenger, $\mathcal{R}$ chooses a bit $b' \xleftarrow{R} \{0, 1\}$, computes a commitment $c_{b'}$ on the message $m_{b'}$ using the string r, then, outputs $\mu = (c_{b'}, e_b, \Sigma.\mathtt{sign}_{\Sigma.sk}(c_{b'}))$ as a challenge signature to $\mathcal{A}$.

Two cases: either μ is a valid confirmer signature on $m_{b'}$ (if $b = b'$), or it is not a valid signature on either m_0 or m_1. However, since the used commitment is statistically hiding, i.e. $c_{b'}$ reveals no information about $m_{b'}$, then μ is conform to a challenge signature in a real INV-CMA game.

[Post challenge phase] $\mathcal{A}$ continues to issue queries and $\mathcal{R}$ continues to handle them as before. Note that at this stage, $\mathcal{R}$ cannot request her PCA oracle on $(e_b, r \| m_i \| \Sigma.pk)$, $i \in \{0, 1\}$. $\mathcal{R}$ would need to make such a query if she gets a verification (conversion) query on a signature $(c_i, e_b, \sigma_i) \neq \mu$ and the message $m_i \in \{m_0, m_1\}$. $\mathcal{R}$ will respond to such a query by running the denial protocol (output $\perp$). This simulation differs from the real algorithm when (c_i, e_b, σ_i) is valid on m_i. Again, such a scenario won't happen with probability at least $(1 - \epsilon')^{q_v+q_{sc}}$, because the query would form a strong existential forgery on the digital signature scheme underlying the construction.

[Final output] The rest of the proof follows in a straightforward way. Let b_a be the bit output by $\mathcal{A}$. $\mathcal{R}$ will output b' to her challenger in case $b' = b_a$ and $1 - b'$ otherwise.

The advantage of $\mathcal{A}$ in such an attack is defined by

$$\epsilon = \mathsf{Adv}(\mathcal{A}) = \left| \Pr[b_a = b' | b = b'] - \frac{1}{2} \right|.$$

We assume again without loss of generality that $\epsilon = \Pr[b_a = b' | b = b'] - \frac{1}{2}$. The advantage of $\mathcal{R}$ is by definition the product $p_{\mathsf{sim}} \cdot p_{\mathsf{chal}}$, where p_{sim} is the probability of providing a simulation indistinguishable from that in a real attack; it is equal to $[(1 - \epsilon_b) \cdot (1 - \epsilon')]^{q_v + q_{sc}}$. Whereas p_{chal} is the probability that $\mathcal{R}$ solves her challenge provided the simulation is correct:

$$\begin{aligned} p_{\mathsf{chal}} &= \left[\Pr[b' = b_a, b = b'] + \Pr[b' \neq b_a, b \neq b'] - \frac{1}{2} \right] \\ &= \frac{1}{2} \left[\Pr[b' = b_a | b = b'] + \Pr[b' \neq b_a | b \neq b'] - 1 \right] \\ &= \left[\frac{1}{2} (\epsilon + \frac{1}{2} + \frac{1}{2} - 1) \right] \\ &= \frac{\epsilon}{2} \end{aligned}$$

Actually, $\Pr[b \neq b'] = \Pr[b = b'] = \frac{1}{2}$ as $b' \overset{R}{\leftarrow} \{0, 1\}$. Moreover, if $b \neq b'$, then the probability that $\mathcal{A}$ answers $1 - b'$ is $\frac{1}{2}$ (since μ is invalid on either m_0 or m_1).

□

References

Gentry C, Molnar D, Ramzan Z (2005) Efficient designated confirmer signatures without random Oracles or general zero-knowledge proofs. In: Roy B (ed) Advances in cryptology - ASIACRYPT 2005. LNCS, vol 3788. Springer, Heidelberg, pp 662–681

Wang G, Baek J, Wong DS, Bao F (2007) On the generic and efficient constructions of secure designated confirmer signatures. In: Okamoto T, Wang X (eds) PKC 2007. LNCS, vol 4450. Springer, Heidelberg, pp 43–60

Chapter 6
CtEtS: An Efficient Variant of CtEaS

Abstract The CtEaS paradigm suffers an intrinsic weakness consisting in the possibility of producing a confirmer signature without knowledge of the signing key. This makes the paradigm rest on strong encryption (PCA secure), and rules out consequently homomorphic encryption which is known for propping up verification. In this chapter, we annihilate this weakness and demonstrate the efficiency of the resulting construction by describing many concrete instantiations. Our modification applies only to confirmer signatures (see Chap. 7 for the details). We further shed light on a special instance of CtEaS, namely Encrypt_then_Sign (EtS), which can be very useful in situations where a trusted party is available.

6.1 Commit_then_Encrypt_then_Sign: CtEtS

We remedy the strong forgeability problem of CtEaS by using the same trick applied to StE, namely bind the used digital signature to the corresponding confirmer signature. This is achieved by producing a digital signature on both the commitment and the encryption of the random string used to generate it. In this way, the attack mentioned in Fact 5.1 no longer applies since an adversary would need to produce a digital signature on the commitment and the re-encryption of the random string used in it. Note that such a fix already appears in the construction of Gentry et al. (2005), however, it was not exploitable as invisibility was considered in the insider model.

6.1.1 The Construction

Consider the following base primitives:

- **A digital signature scheme** Σ given by (1) Σ.`keygen` which generates a key pair ($\Sigma.sk$, $\Sigma.pk$) (2) Σ.`sign` (3) Σ.`verify`.
- **A public-key encryption scheme** Γ described by (1) Γ.`keygen` that generates the key pair ($\Gamma.sk$,$\Gamma.pk$) (2) Γ.`encrypt` (3) Γ.`decrypt`.
- **A commitment scheme** Ω given by (1) Ω.`commit` (2) Ω.`open`.

L. El Aimani, *Verifiable Composition of Signature and Encryption*,
https://doi.org/10.1007/978-3-319-68112-2_6

setup(1^κ)	: Σ.setup(1^κ) ; Γ.setup(1^κ) ; Ω.setup(1^κ)
keygen$_S$(1^κ)	: Σ.keygen(1^κ)
keygen$_C$(1^κ)	: Γ.keygen(1^κ)
sign(m)	: $c \leftarrow \Omega$.commit(m,r) ; $e \leftarrow \Gamma$.encrypt$_{\{\Gamma.pk,coins_e\}}(r)$
	$\sigma \leftarrow \Sigma$.sign$_{\Sigma.sk}(e\|c)$; return (c,e,σ)
sconfirm($\{c,e,\sigma\}$,m)	: ZKP$\{(r,coins_e): c = \Omega.\mathtt{commit}(m,r) \wedge e = \Gamma.\mathtt{encrypt}_{\{\Gamma.pk,coins_e\}}(r)\}$
confirm($\{c,e,\sigma\}$,m)	: ZKP$\{(r,\Gamma.sk): c = \Omega.\mathtt{commit}(m,r) \wedge r = \Gamma.\mathtt{decrypt}_{\Gamma.sk}(e)\}$
deny($\{c,e,\sigma\}$,m)	: ZKP$\{(r,\Gamma.sk): c \neq \Omega.\mathtt{commit}(m,r) \wedge r = \Gamma.\mathtt{decrypt}_{\Gamma.sk}(e)\}$
convert($\{c,e,\sigma\}$,m)	: $r \leftarrow \Gamma$.decrypt$_{\Gamma.sk}(c)$; $b \leftarrow (c = \Omega.\mathtt{commit}(m,r))$
	if $b = 0$ return $(\perp)$ else return (r,c,σ)

Fig. 6.1 The CtEtS paradigm

We assume that Γ produces ciphertexts of length exactly a certain κ. As a result, the first bit of c will always be at the $(\kappa + 1)$th position in $e\|c$, where e is an encryption produced by Γ. The construction from the CtEtS paradigm is given in Fig. 6.1. Needless to say that sconfirm, confirm, and deny are only carried out when the signature σ on $e\|c$ is valid, otherwise the confirmer signature $\mu = (c, e, \sigma)$ is clearly deemed invalid w.r.t. m.

It is clear that this new construction looses the parallelism of the original one, i.e. encryption and signature can no longer be carried out in parallel, however, it has the advantage of resting on cheaper encryption as we will show in the following. Moreover, it still preserves the merit of the CtEaS paradigm, namely the possibility to instantiate with *any* digital signature.[1]

6.1.2 Security Analysis

Theorem 6.1 (Unforgeability of CtEtS) *Given* $(t, q_s) \in \mathbb{N}^2$ *and* $(\epsilon, \epsilon') \in [0, 1]^2$, *the construction depicted above is* $(t, \epsilon \cdot (1 - \epsilon')^{q_s}, q_s)$-EUF-CMA *secure if it uses a* (t, ϵ')*-binding commitment and a* (t, ϵ, q_s)-EUF-CMA *secure digital signature scheme.*

Proof Let $\mathcal{A}$ be an EUF-CMA attacker against the construction. We construct an EUF-CMA attacker $\mathcal{R}$ against the underlying digital signature scheme as follows.

$\mathcal{R}$ gets the parameters of the digital signature from her attacker, and chooses suitable encryption and commitment schemes. After she gets the confirmer's key pair from $\mathcal{A}$, $\mathcal{R}$ can perfectly simulate signature queries using her own challenger. At some point, $\mathcal{A}$ will output a forgery $\mu^\star = (c^\star, e^\star, \sigma^\star)$ on some message $m^\star$,

[1]Practical realizations from the StE paradigm need to use digital signatures from a special class that we specified in Definition 4.2.

which was never queried before for signature. By definition, $\sigma^\star$ is a valid digital signature on $e^\star \| c^\star$. Suppose there exists $1 \le i \le q_s$ such that $e^\star \| c^\star = e_i \| c_i$ where $\mu_i = (c_i, e_i, \sigma_i)$ was the output confirmer signature on the query m_i. Due to the special way the strings $e_i \| c_i$ are created, this implies $(e_i, c_i) = (e^\star, c^\star)$. With probability at least $(1 - \epsilon')$, we have $m_i = m^\star$ (the commitment is (t, ϵ')-binding), which is a contradiction. Therefore, $e^\star \| c^\star$ was not queried by $\mathcal{R}$ for a digital signature with probability at least $(1 - \epsilon')^{q_s}$. $\mathcal{R}$ outputs to her challenger the EUF-CMA forgery $\sigma^\star$ and $e^\star \| c^\star$. □

For the invisibility proof, we first need this lemma:

Lemma 6.1 *Let Ω and Γ be a commitment and a public-key encryption scheme respectively. We consider the following game between an adversary $\mathcal{A}$ and his challenger $\mathcal{R}$:*

1. *$\mathcal{R}$ invokes the algorithms Γ.keygen(1^κ) to generate (pk, sk), where κ is a security parameter.*
2. *$\mathcal{A}$ outputs two messages m_0 and m_1, such that $m_0 \neq m_1$, to his challenger.*
3. *$\mathcal{R}$ generates two random strings r_0 and r_1 such that $r_0 \neq r_1$. Next, she chooses two bits $b, b' \xleftarrow{R} \{0, 1\}$ uniformly at random. Finally, she outputs to $\mathcal{A}$ $c_b = \Omega.\texttt{commit}(m_b, r_{1-b'})$ and $e_{b'} = \Gamma.\texttt{encrypt}_{pk}(r_{b'})$.*
4. *$\mathcal{A}$ outputs a bit b_a representing his guess of c_b not being the commitment of m_b using the nonce $\Gamma.\texttt{decrypt}(e_{b'})$. $\mathcal{A}$ wins the game if $b_a \neq b$, and we define his advantage as*

$$\mathsf{Adv}(\mathcal{A}) = \left| \Pr[b \neq b_a] - \frac{1}{2} \right|,$$

where the probability is taken over the random tosses of both $\mathcal{A}$ and $\mathcal{R}$.

If Ω is injective, (t, ϵ_b)-binding, and (t, ϵ_h)-hiding, then $\mathsf{Adv}(\mathcal{A})$ in the above game is at most $\frac{\epsilon_h}{1-\epsilon_b}$.

Proof Let ϵ be the advantage of $\mathcal{A}$ in the game above. We will construct an adversary $\mathcal{R}$ which breaks the hiding property of the used commitment with advantage $\epsilon \cdot (1 - \epsilon_b)$.

- $\mathcal{R}$ gets from $\mathcal{A}$ the messages m_0, m_1, and forwards them to her own challenger.
- $\mathcal{R}$ receives from her challenger the commitment $c_b = \Omega.\texttt{commit}(m_b, r)$ for some $b \xleftarrow{R} \{0, 1\}$ and some nonce r.
- $\mathcal{R}$ generates a nonce r' and outputs to $\mathcal{A}$ c_b and $e = \Gamma.\texttt{encrypt}_{pk}(r')$.
- When $\mathcal{A}$ outputs a bit b_a, $\mathcal{R}$ outputs to her challenger $1 - b_a$.

If $\mathcal{A}$ can by some means get hold of r', then he can compute $c_i = \Omega.\texttt{commit}(m_i, r')$, $i = 0, 1$. Since Ω is injective and binding, then $c_b \neq \Omega.\texttt{commit}(m_b, r')$ and $c_b \neq \Omega.\texttt{commit}(m_{1-b}, r')$ respectively, i.e. $c_b \notin \{c_0, c_1\}$. Thus, $\mathcal{A}$ will get no information on the message underlying c_b even if he manages to invert e.

We have by definition:

$$\begin{aligned}\mathsf{Adv}(\mathcal{R}) &= (1-\epsilon_b)\left|\Pr[1-b_a=b]-\frac{1}{2}\right|\\ &= (1-\epsilon_b)\left|\Pr[b_a\neq b]-\frac{1}{2}\right|\\ &= \epsilon\cdot(1-\epsilon_b)\end{aligned}$$

□

Remark 6.1 Note that the above lemma holds true regardless of the used encryption Γ. For instance, it can be used with encryption schemes that do not require any kind of security.

Theorem 6.2 (Invisibility of CtEtS) *Given* $(t, q_s, q_v, q_{sc}) \in \mathbb{N}^4$ *and* $(\epsilon, \epsilon', \epsilon_h, \epsilon_b) \in [0,1]^4$, *the construction depicted above is* $(t, \epsilon, q_s, q_v, q_{sc})$-INV-CMA *secure if it uses a* (t, ϵ', q_s)-SEUF-CMA *secure digital signature, an injective,* (t, ϵ_b)*-binding, and* (t, ϵ_h)*-hiding commitment, and a* $(t + q_s(q_v + q_{sc}), \frac{1}{2}(\epsilon + \frac{\epsilon_h}{1-\epsilon_b})(1 - \frac{\epsilon_h}{1-\epsilon_b}) \cdot [(1-\epsilon_b)\cdot(1-\epsilon')]^{q_v+q_{sc}})$-IND-CPA *secure encryption scheme.*

Proof (Sketch) Let Σ, Γ, and Ω be the signature, encryption, and commitment schemes underlying the construction resp. Let further $\mathcal{R}$ be the reduction using the invisibility attacker $\mathcal{A}$ in order to break Γ.

$\mathcal{R}$ gets the public key of Γ from her challenger. She further generates the parameters of Σ (for instance ($\Sigma.sk$,$\Sigma.pk$)) and of Ω.

Pre-challenge phase. Simulation of `sign` and `sconfirm` queries is done as dictated by the standard algorithm/protocol, with the exception of maintaining a list $\mathcal{L}$ of the strings used to produce commitments on the queried messages in addition to their encryptions.

For a verification (conversion) query, $\mathcal{R}$ looks up the list $\mathcal{L}$ for the decryption of the second component of the signature; if it is found, $\mathcal{R}$ simulates the confirmation protocol (issues the converted signature in case of a conversion query), otherwise she simulates the denial protocol (issues the symbol $\perp$ in case of a conversion). The difference between this simulation and the real execution of the algorithm manifests when a queried signature, say (c_i, e_i, σ_i), is valid, on the queried message m_i, but e_i is not present on the list. We distinguish two cases, either the underlying message m_i has been queried previously or not. In the latter case, such a signature would correspond to an existential forgery on the construction, thus, to an existential forgery on Σ or to breaking the binding property of Ω. In the former case, let (c_j, e_j, σ_j) be the output signature to $\mathcal{A}$ on the message m_i. We have $e_i \| c_i \neq e_j \| c_j$ since $e_i \neq e_j$, and both e_i and e_j are the κ-bit prefixes of $e_i \| c_i$ and $e_j \| c_j$ resp. We conclude that the adversary would have to compute a digital signature on a string for which he had never obtained a signature. Thus, the query would lead to an existential forgery on Σ.

Bottom line is, the probability that the provided simulation does not deviate from the real execution is at least $[(1-\epsilon') \cdot (1-\epsilon_b)]^{q_v+q_{sc}}$.

Challenge phase. At some point, $\mathcal{A}$ outputs two messages m_0, m_1 to $\mathcal{R}$. The latter chooses two different random strings r_0 and r_1 and hands them to her challenger. $\mathcal{R}$ receives then a ciphertext $e_{b'}$, encryption of $r_{b'}$, for some $b' \in \{0,1\}$. To answer her challenger, $\mathcal{R}$ computes a commitment c_b on the message m_b for some $b \overset{R}{\leftarrow} \{0,1\}$ using the string r_b, then outputs $\mu = (c_b, e_{b'}, \Sigma.\mathtt{sign}_{\Sigma.sk}(e_{b'} \| c_b))$ as a challenge confirmer signature to $\mathcal{A}$. Two cases: either μ is a valid confirmer signature on m_b (if $b' = b$), or it is not a valid signature on either m_0 or m_1. $\mathcal{A}$ cannot tell the difference between the provided challenge and that in a real attack with probability at least $1 - \frac{\epsilon_h}{1-\epsilon_b}$ according to Lemma 6.1.

Post-challenge phase. $\mathcal{A}$ continues to issue queries and $\mathcal{R}$ continues to handle them as before. Note that in this phase, $\mathcal{R}$ might get a verification (conversion) query on a signature $(c_b, e'_b, -) \neq \mu$ and the message m_b. $\mathcal{R}$ will respond to such a query by running the denial protocol (output $\perp$). This simulation differs from the real algorithm when $(c_b, e'_b, -)$ is valid on m_b. Again, such a scenario won't happen with probability at least $[(1-\epsilon') \cdot (1-\epsilon_b)]^{q_v+q_{sc}}$, because the query would form a strong existential forgery on Σ.

Final output. Let b_a be the bit output by $\mathcal{A}$. $\mathcal{R}$ will output b to her challenger in case $b = b_a$ and $1-b$ otherwise.

The advantage of $\mathcal{A}$ in such an attack is defined by

$$\epsilon = \mathsf{Adv}(\mathcal{A}) = \left| \Pr[b_a = b | b' = b] - \frac{1}{2} \right|$$

We assume again without loss of generality that $\epsilon = \Pr[b_a = b | b' = b] - \frac{1}{2}$. The advantage of $\mathcal{R}$ is then given by the product $p_{\mathsf{sim}} \cdot p_{\mathsf{chal}}$, where p_{sim} is the probability of providing a simulation indistinguishable from that in a real attack; it is equal to $(1 - \frac{\epsilon_h}{1-\epsilon_b}) \cdot [(1-\epsilon_b) \cdot (1-\epsilon')]^{q_v+q_{sc}}$. Whereas p_{chal} is the probability that $\mathcal{R}$ solves her challenge provided the simulation is correct:

$$\begin{aligned}
p_{\mathsf{chal}} &= \left[\Pr[b = b_a, b' = b] + \Pr[b \neq b_a, b' \neq b] - \frac{1}{2} \right] \\
&= \frac{1}{2} \left[\Pr[b = b_a | b' = b] + \Pr[b \neq b_a | b' \neq b] - 1 \right] \\
&= \frac{1}{2} \left[(\epsilon + \frac{1}{2}) + (\frac{\epsilon_h}{1-\epsilon_b} + \frac{1}{2}) - 1 \right] \\
&= \frac{1}{2} (\epsilon + \frac{\epsilon_h}{1-\epsilon_b})
\end{aligned}$$

In fact, $\Pr[b' \neq b] = \Pr[b' = b] = \frac{1}{2}$ as $b \xleftarrow{R} \{0, 1\}$. Moreover, if $b' \neq b$, then the probability that $\mathcal{A}$ answers $1 - b$ is $\frac{1}{2}$ greater than the advantage of the adversary in the game defined in Lemma 6.1, namely $\frac{\epsilon_h}{1-\epsilon_b}$. □

6.1.3 *Practical Instantiations*

6.1.3.1 The Class $\mathbb{C}$ of Commitments

Definition 6.1 (Class $\mathbb{C}$ of Commitments) $\mathbb{C}$ is the set of all commitment schemes for which there exists an algorithm `compute` that inputs the commitment key *pk*, the message *m* and the commitment *c* on *m*, and computes a description of a map $f : (\mathbb{G}, *) \to (\mathbb{H}, \circ_c)$ where:

- $(\mathbb{G}, *)$ is a group and $\mathbb{H}$ is a set equipped with the binary operation $\circ_c$,
- $\forall r, r' \in \mathbb{G}: f(r * r') = f(r) \circ_c f(r')$.

and an $I \in \mathbb{H}$, such that $f(r) = I$, where r is the opening value (decommitment) of c w.r.t. m.

It is easy to check that Pedersen's commitment scheme outlined in Fig. 1.3 is in this class. Actually, most commitment schemes have this built-in property because it is often the case that the committer wants to prove efficiently that a commitment is produced on some message. This is possible if the function f is homomorphic as shown in Fig. 6.2.

Theorem 6.3 *The protocol depicted in Fig. 6.2 is a perfect zero-knowledge proof of knowledge of preimages of the function f described in Definition 6.1.* □

6.1.3.2 Confirmation/Denial Protocols

The CtEtS has the merit of supporting *any* digital signature scheme as a building block. In this way, confirmation (denial) of a confirmer signature on a certain message consists only in proving knowledge of the decryption of a given ciphertext,

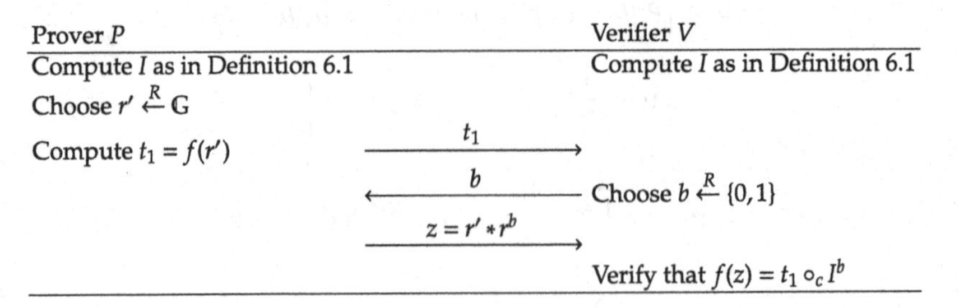

Prover P
Verifier V
Compute I as in Definition 6.1
Compute I as in Definition 6.1
Choose $r' \xleftarrow{R} \mathbb{G}$
t_1
Compute $t_1 = f(r')$
b
Choose $b \xleftarrow{R} \{0,1\}$
$z = r' * r^b$
Verify that $f(z) = t_1 \circ_c I^b$

Fig. 6.2 Proof of membership to the language $\{c: c = \texttt{commit}(m, r)\}$ **Common input:** (c, m) and **Private input** : r

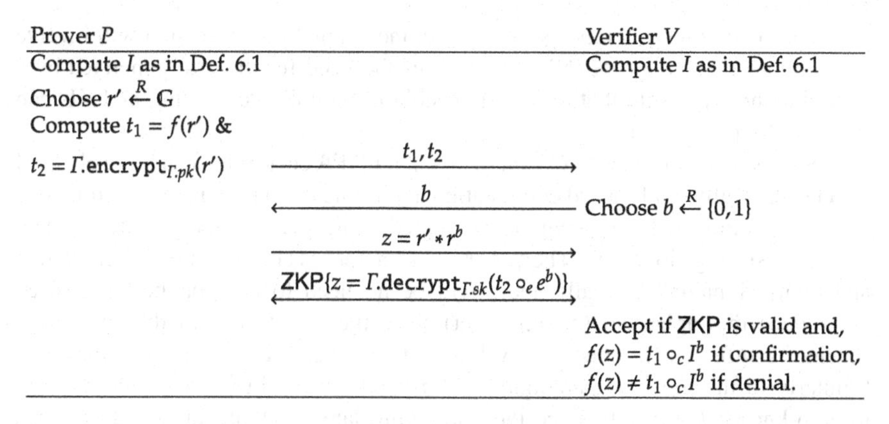

Prover P
Verifier V
Compute I as in Def. 6.1
Compute I as in Def. 6.1
Choose $r' \xleftarrow{R} \mathbb{G}$
Compute $t_1 = f(r')$ &
$t_2 = \Gamma.\mathtt{encrypt}_{\Gamma.pk}(r')$
t_1, t_2
b
Choose $b \xleftarrow{R} \{0,1\}$
$z = r' * r^b$
ZKP$\{z = \Gamma.\mathtt{decrypt}_{\Gamma.sk}(t_2 \circ_e e^b)\}$
Accept if ZKP is valid and,
$f(z) = t_1 \circ_c I^b$ if confirmation,
$f(z) \neq t_1 \circ_c I^b$ if denial.

Fig. 6.3 Confirmation/denial protocol for CtEtS: PoK$\{r: r = \Gamma.\mathtt{decrypt}(e) \wedge c = (\neq)\ \Omega.\mathtt{commit}(m, r)\}$ **Common input:** $(e, c, m, \Gamma.pk, \Omega.pk)$ and **Private input:** $\Gamma.sk$ or randomness encrypting r in e

and that this decryption is (is not) the opening value of a given commitment on the message. More specifically, the confirmation/denial protocols for CtEtS, when the encryption Γ belongs to the class $\mathbb{E}$ and the commitment Ω belongs to the class $\mathbb{C}$, are depicted in Fig. 6.3.

Theorem 6.4 *Let Ω and Γ be commitment and encryption schemes from the classes $\mathbb{C}$ and $\mathbb{E}$ resp. The confirmation protocol depicted in Fig. 6.3 is a perfect zero-knowledge proof of knowledge.*

Theorem 6.5 *The denial protocol depicted in Fig. 6.3, for $\Omega \in \mathbb{C}$ and $\Gamma \in \mathbb{E}$, is a proof of knowledge with computational zero-knowledge if Γ is* IND-CPA *secure.*

The proofs are similar to those of Theorem 4.5 and Theorem 4.6 respectively. □

CtEtS achieves better performances than the original technique in terms of bandwidth and cost while supporting any signature scheme in its base components. Moreover, it accepts many efficient instantiations (if the used commitment and encryption belong to the already mentioned classes) as its confirmation/denial protocols no longer rely on general proofs of NP statements. Besides, the techniques presented in Chap. 4 for the new StE, still apply to CtEtS to reach concurrent zero-knowledge and online non-transferability, in addition to reducing the soundness error of the verification protocols.

6.2 The "Encrypt_then_Sign" (EtS) Paradigm

This paradigm was first used in the context of signcryption (An et al. 2002) that does not support signature extraction. We can adapt it to the case of convertible confirmer signatures by requiring a "trusted authority" TA that runs the `setup` algorithm and generates a common reference string crs. In fact, signature conversion will involve a non-interactive ZK proof (NIZK), and thus the need for the crs (generated by a trusted authority). Note that a TA is plausible in a public-key setting as PKIs can successfully play its role.

EtS is a special instance of CtEtS. In fact, IND-CPA encryption can be easily used to get statistically binding and computationally hiding commitments. Therefore, one can first commit to the message to be signed using the encryption scheme, then sign the resulting ciphertext. The confirmer signature is composed of the ciphertext and of its signature. Actually, there will be no need to encrypt the string used to produce the ciphertext (commitment) since the private key of the encryption scheme is sufficient to check the validity of a ciphertext w.r.t. a given message. Similarly to the previous paradigms, `sconfirm`, `confirm`, and `deny` amount to zero-knowledge proofs since the underlying languages are in NP. It is worth noting that the aforementioned protocols are only carried out when the signature σ on the ciphertext c is valid, otherwise the confirmer signature $\mu = (c, \sigma)$ is obviously deemed invalid w.r.t. m. Finally, `convert` outputs (in case of a valid confirmer signature on m) a non-interactive zero-knowledge (NIZK) proof that m is the decryption of c; such a proof is feasible since the underlying statement is in NP (Blum et al. 1988; Goldreich et al. 1986).

We describe in Fig. 6.4 confirmer signatures from EtS, where the base components are the digital signature Σ and the public-key encryption Γ.

6.2.1 *Security Analysis*

We first note that completeness, soundness, and (offline) non-transferability of the verification protocols are ensured by the use of ZK proofs. Next, we prove that

CS.setup(1^κ)	: Σ.setup(1^κ) ; Γ.setup(1^κ) ; crs $\leftarrow$ TA.setup(1^κ)
CS.keygen$_S$(1^κ)	: Σ.keygen(1^κ)
CS.keygen$_C$(1^κ)	: Γ.keygen(1^κ)
CS.sign(m)	: $c \leftarrow \Gamma.\mathsf{encrypt}_{\{\Gamma.pk, coins_c\}}(m)$; $\sigma \leftarrow \Sigma.\mathsf{sign}_{\Sigma.sk}(c)$ return (c,σ)
CS.sconfirm({c,σ},m)	: $\mathsf{ZKP}\left\{coins_c \colon c = \Gamma.\mathsf{encrypt}_{\{\Gamma.pk, coins_c\}}(m)\right\}$
CS.confirm({c,σ},m)	: $\mathsf{ZKP}\left\{\Gamma.sk \colon m = \Gamma.\mathsf{decrypt}_{\Gamma.sk}(c)\right\}$
CS.deny({c,σ},m)	: $\mathsf{ZKP}\left\{\Gamma.sk \colon m \neq \Gamma.\mathsf{decrypt}_{\Gamma.sk}(c)\right\}$
CS.convert({c,σ},m)	: $\pi \leftarrow \mathsf{NIZK}\left\{m = \Gamma.\mathsf{decrypt}_{\Gamma.sk}(c)\right\}$; return (π,c,σ)
CS.verifyconverted({π,c,σ},m)	: NIZK.verify(crs,π) ; $\Sigma.\mathsf{verify}_{\Sigma.pk}(\sigma,c)$

Fig. 6.4 The EtS paradigm

the construction resists existential forgeries and is invisible if the underlying digital signature and encryption are SEUF-CMA and IND-CPA secure resp.

Theorem 6.6 (Unforgeability of EtS) *Given* $(t, q_s) \in \mathbb{N}^2$ *and* $\epsilon \in [0, 1]$*, EtS is* (t, ϵ, q_s)-EUF-CMA *secure if the underlying digital signature is also* (t, ϵ, q_s)-EUF-CMA *secure.*

Proof The adversary $\mathcal{R}$ against the signature underlying the construction gets the parameters of the digital signature he is trying to attack from his challenger. Then, he chooses a suitable encryption. Simulation of signatures is done by first encrypting the message to be signed, then requesting his challenger for a signature on the result encryption.

At some time, the adversary $\mathcal{A}$ against the construction will output a forgery $(c^\star, \sigma^\star)$ on a message $m^\star$, that was never queried before. $\sigma^\star$ is by definition a digital signature on $c^\star$. The last item was never queried by $\mathcal{R}$ for digital signature, since otherwise $m^\star$ would have been queried before. We conclude that $(c^\star, \sigma^\star)$ is a valid forgery on the digital signature scheme. □

Theorem 6.7 (Invisibility of EtS) *Given* $(t, q_s, q_v, q_{sc}) \in \mathbb{N}^4$ *and* $(\epsilon, \epsilon') \in [0, 1]^2$*, the construction given above is* $(t, \epsilon, q_s, q_v, q_{sc})$-INV-CMA *secure if it uses a* (t, ϵ', q_s)-SEUF-CMA *secure digital signature and a* $(t + q_s(q_v + q_{sc}), \epsilon(1 - \epsilon')^{q_v+q_{sc}})$-IND-CPA *secure encryption scheme.*

Proof Let $\mathcal{A}$ be the invisibility adversary against the construction, we construct an IND-CPA adversary $\mathcal{R}$ against the underlying encryption scheme as follows.

$\mathcal{R}$ gets the parameters of the target encryption scheme from his challenger, and chooses a suitable digital signature scheme. Signature queries are simulated as dictated by the real algorithm, with the exception of maintaining a list $\mathcal{L}$ of records that consist of the queried message, its encryption, the randomness used to produce the encryption, and finally the digital signature on the encryption. $\mathcal{R}$ can confirm a just generated signature with the knowledge of the randomness used in the encryption.

For a verification query (c_i, σ_i) on m_i, $\mathcal{R}$ checks $\mathcal{L}$ (after checking the validity of σ_i on c_i), if the record $R_i = (m_i, c_i, -, -)$ appears in the list, then he will simulate a proof that c_i decrypts in m_i. Otherwise, he will simulate a proof of the inequality of the decryption of c_i and m_i.

For a conversion query, $\mathcal{R}$ proceeds as in a verification query with the exception of providing the non-interactive variant of the proof he would issue if the signature is valid (using crs), and the symbol $\perp$ otherwise.

This simulation differs from the real one when the queried signature (c_i, σ_i) is valid on m_i however c_i does not appear in the list. We distinguish two cases, either m_i was not queried before for signature, which corresponds to an existential forgery on EtS, and thus on the underling signature scheme. Or, the queried signature is on a message that has been queried before, which corresponds to an existential forgery on the underlying signature scheme. Since the signature scheme underlying the construction is (t, ϵ', q_s)-SEUF-CMA secure, this scenario does not happen with probability at least $(1 - \epsilon')^{q_v+q_{sc}}$.

At some point, $\mathcal{A}$ produces two messages m_0, m_1. $\mathcal{R}$ will forward the same messages to his challenger and obtain a ciphertext c, encryption of m_b for some $b \xleftarrow{R} \{0,1\}$. $\mathcal{R}$ will produce a digital signature σ on c and gives the result in addition to c to $\mathcal{A}$ as a challenge confirmer signature. It easy to see that $\mathcal{A}$'s answer is sufficient for $\mathcal{R}$ to conclude. Note that after the challenge phase, $\mathcal{A}$ is allowed to issue verification and conversion queries and $\mathcal{R}$ can handle them as previously. There is however the possibility for $\mathcal{A}$ of issuing a verification (conversion) query of the type $(c, -) \neq (c, \sigma)$ on m_b. $\mathcal{R}$ will respond to such a query by issuing the denial protocol (symbol $\perp$). The probability that this answer does not differ from the output of the real algorithm is at least $(1 - \epsilon')^{q_v + q_{sc}}$ as the signature scheme underlying the construction is (t, ϵ', q_s)-SEUF-CMA secure by assumption. □

Remark 6.2 Note that the IND-CPA requirement on the encryption scheme is also necessary. In fact, an invisibility adversary against the construction can easily use an IND-CPA adversary against the underlying encryption scheme in order to solve his invisibility challenge.

6.2.2 Confirmation/Denial Protocols

Confirmation in EtS amounts to a proof of correctness of a decryption (i.e. a given ciphertext encrypts a given message). This is in general easy since in most encryption schemes, one can define, given a ciphertext c and its underlying plaintext m, two homomorphic maps f and g, and two quantities I and J such that $f(r) = I$ and $g(sk) = J$, where r is the randomness used to encrypt m in c, and sk is the private key of the encryption scheme. Examples of such encryptions include El Gamal (1985), Boneh et al. (2004), Paillier (1999), Cramer and Shoup (2003) and Camenisch and Shoup (2003). The confirmation protocol run by the signer (confirmer) in this case will be reduced to a proof of knowledge of a preimage of I (resp. J) by the function f (resp. g), for which we provided an efficient proof in Fig. 4.2.

The denial protocol is not always straightforward. In most discrete-logarithm-based encryptions, this protocol amounts to a proof of inequality of discrete logarithms as in El Gamal (1985), Boneh et al. (2004), and Cramer and Shoup (2003). In case the encryption scheme belongs to the class $\mathbb{E}$ defined earlier, Fig. 6.5 provides an efficient proof that c encrypts some $\tilde{m}$ which is different from m. In the protocol depicted in this figure, f denotes an arbitrary *homomorphic one-way* map:

$$f(m * m') = f(m) \circ f(m')$$

Similarly, the above denial protocol can be shown to be a proof of knowledge with computational ZK if Γ is IND-CPA secure.

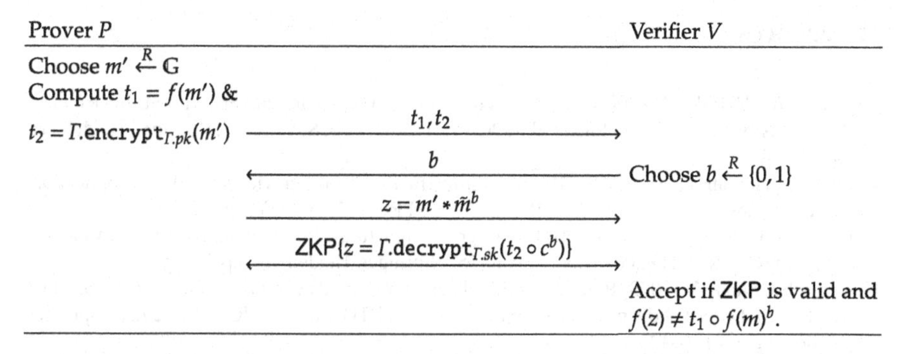

Fig. 6.5 Denial protocol in EtS. $\mathsf{PoK}\{\tilde{m}: \tilde{m} = \Gamma.\mathtt{decrypt}(c) \wedge \tilde{m} \neq m\}$ **Common input:** $(m, c, \Gamma.pk)$ and **Private input:** $\tilde{m}$ and $\Gamma.sk$ or randomness encrypting $\tilde{m}$ in c

6.2.3 Selective Conversion

Selective conversion in confirmer signatures from EtS consists in providing the non-interactive variant of the confirmation protocol. We note in this paragraph few solutions to achieve this goal.

The case of fully decryptable encryption schemes That is encryption schemes where decryption leads to the randomness used to produce the ciphertext. In this case, selective conversion can simply be achieved by releasing the randomness used to generate the ciphertext. Examples of encryption schemes from this class include Paillier's encryption (Paillier 1999): the scheme operates on messages in $\mathbb{Z}_N$, where $N = pq$ is a safe RSA modulus. Encryption of a message m is done by picking a random $r \in_R \mathbb{Z}_N^\times$ and then computing the ciphertext $c \equiv r^N(1 + mN) \bmod N^2$. Decryption of a ciphertext c is done by raising it to $\lambda = \mathrm{lcm}(p-1, q-1)$ to find m. It is easy to see that recovering r, once m is computed, amounts to an extraction of the Nth root of $\frac{c}{1+mN}$.

Damgård et al. (2006)'s solution Recall from Sect. 1.4.4 that this solution transforms a three-move interactive ZK protocol P with linear answer to a non-interactive ZK one (NIZK) using a homomorphic encryption scheme in a registered key model. Thus, EtS accepts an efficient instantiation if proving the correctness of the decryption (of the considered encryption scheme) amounts to a proof of equality of two discrete logarithms, e.g. El Gamal (1985), Boneh et al. (2004), and Cramer and Shoup (2003).

Groth and Sahai (2008)'s solution This technique is applicable in general for encryption schemes where the encryption/decryption algorithms perform only group or pairing (if bilinear groups are involved) operations on the randomness or the private key.

Lindell (2014)'s solution This Fiat-Shamir like transform turns any Σ protocol for a relation R into a NIZK proof for the associated language L_R, while alleviating the use of the random oracle that afflicted the original method. The concrete computational complexity of this technique is slightly higher than that of the original Fiat-Shamir transform.

References

An JH, Dodis Y, Rabin T (2002) On the security of joint signature and encryption. In: Knudsen LR (ed) Advances in cryptology - EUROCRYPT 2002. LNCS, vol 2332. Springer, Heidelberg, pp 83–107

Blum M, Feldman P, Micali S (1988) Non-interactive zero-knowledge and its applications (extended abstract). In: Simon J (ed) STOC. ACM Press, New York, pp 103–112

Boneh D, Boyen X, Shacham H (2004) Short group signatures. In: Franklin MK (ed) Advances in cryptology - CRYPTO 2004. LNCS, vol 3152. Springer, Heidelberg, pp 41–55

Camenisch J, Shoup V (2003) Practical verifiable encryption and decryption of discrete logarithms. In: Boneh D (ed) Advances in cryptology - CRYPTO 2003. LNCS, vol 2729. Springer, Heidelberg, pp 126–144

Cramer R, Shoup V (2003) Design and analysis of practical public-key encryption schemes secure against adaptive chosen ciphertext attack. SIAM J Comput 33(1):167–226

Damgård I, Fazio N, Nicolosi A (2006) Non-interactive zero-knowledge from homomorphic encryption. In: Halevi S, Rabin T (eds) TCC 2006. LNCS, vol 3876. Springer, Heidelberg, pp 41–59

El Gamal T (1985) A public key cryptosystem and a signature scheme based on discrete logarithms. IEEE Trans Inf Theory 31:469–472

Gentry C, Molnar D, Ramzan Z (2005) Efficient designated confirmer signatures without random Oracles or general zero-knowledge proofs. In: Roy B (ed) Advances in cryptology - ASIACRYPT 2005. LNCS, vol 3788. Springer, Heidelberg, pp 662–681

Goldreich O, Micali S, Wigderson A (1986) How to prove all NP-statements in zero-knowledge, and a methodology of cryptographic protocol design. In: Odlyzko AM (ed) CRYPTO. LNCS, vol 263. Springer, Heidelberg, pp 171–185

Groth J, Sahai A (2008) Efficient non-interactive proof systems for bilinear groups. In: Smart NP (ed) EUROCRYPT 2008. LNCS, vol 4965. Springer, Heidelberg, pp 415–432

Lindell Y (2014) An efficient transform from sigma protocols to NIZK with a CRS and non-programmable random Oracle. IACR Cryptology ePrint Archive 2014:710

Paillier P (1999) Public-key cryptosystems based on composite degree residuosity classes. In: Stern J (ed) EUROCRYPT. LNCS, vol 1592. Springer, Heidelberg, pp 223–238

Part IV
New Paradigms

Chapter 7
EtStE: A New Paradigm for Verifiable Signcryption

Abstract The new StE or CtEtS paradigms, proposed earlier, proved to provide very efficient confirmer signatures. Unfortunately, when applied to verifiable signcryption, these paradigms fail to give similar results. The reason lies in the fact that encryptions are produced on the message, to be signcrypted, in addition to other strings (signatures or decommitments), which renders verification ineffective. The subject of this chapter is a new paradigm for verifiable signcryption which combines the merits of the classical paradigms while avoiding their drawbacks.

7.1 Shortcomings of the Classical Paradigms

7.1.1 Review of the Classical Paradigms

Let Σ be a digital signature scheme given by Σ.keygen which generates a key pair ($\Sigma.sk$, $\Sigma.pk$), Σ.sign, and Σ.verify. Let furthermore Γ denote a public-key encryption scheme described by Γ.keygen that generates the key pair ($\Gamma.sk$,$\Gamma.pk$), Γ.encrypt, and Γ.decrypt. Finally, let Ω be a commitment scheme given by the algorithms Ω.commit and Ω.open. The most popular paradigms used to devise signcryption schemes from basic primitives are:

- *"Sign_then_Encrypt" (StE)* (An et al. 2002; Chiba et al. 2011; Matsuda et al. 2009). Given a message m, signcrypt first produces a signature σ on the message using $\Sigma.sk$, then encrypts $m\|\sigma$ under $\Gamma.pk$. The result forms the signcryption on m. To unsigncrypt, one first decrypts the signcryption using $\Gamma.sk$ in $m\|\sigma$, then checks the validity of σ, using $\Sigma.pk$, on m. Finally, sigExtract of a valid signcryption $\mu = \Gamma.\texttt{encrypt}(m\|\sigma)$ on m outputs σ.
- *"Encrypt_then_Sign" (EtS)* (An et al. 2002; Matsuda et al. 2009). Given a message m, signcrypt produces an encryption e on m using $\Gamma.pk$, then produces a signature σ on e using $\Sigma.sk$; the signcryption is the pair (e, σ). To unsigncrypt such a signcryption, one first checks the validity of σ w.r.t. e using $\Sigma.pk$, then decrypts e using $\Gamma.sk$ to get m. Finally, sigExtract outputs a non-interactive zero-knowledge (NIZK) proof that m is the decryption of e;

L. El Aimani, *Verifiable Composition of Signature and Encryption*,
https://doi.org/10.1007/978-3-319-68112-2_7

such a proof is possible since the statement in question is in NP (Blum et al. 1988; Goldreich et al. 1986). This paradigm naturally requires the presence of a trusted authority in order to generate the common reference string needed for the NIZK proofs.
- *"Commit_then_Encrypt_and_Sign" (CtEaS)* (An et al. 2002). This construction has the advantage of performing signature and encryption *in parallel* in contrast to the previous sequential compositions. Given a message m, one first produces a commitment c on it using some random nonce r, then encrypts $m\|r$ under $\Gamma.pk$, *and* produces a signature σ on c using $\Sigma.sk$. The signcryption is the triple (c, e, σ). To `unsigncrypt` such a signcryption, one first checks the validity of σ w.r.t. c, then decrypts e to get $m\|r$, and finally checks the validity of the commitment c w.r.t (m, r). `sigExtract` is achieved by releasing the decryption of e, namely $m\|r$.

The proofs of well (mal) formed-ness, namely proveValidity and {`confirm`, `deny`} can be carried out since the underlying languages are in NP and thus accept zero-knowledge proof systems (Goldreich et al. 1986). Finally, it is possible to require a proof in the `sigExtract` algorithms of StE and CtEaS, that the revealed information is indeed a correct decryption of the encryption in question; such a proof is again possible to issue since the corresponding statement is in NP.

Theorem 7.1 (Indistinguishability of StE and CtEaS) *Consider the security notions obtained from pairing a security goal* GOAL $\in$ {OW, IND, NM} *and an attack model* ATK $\in$ {CPA, PCA, CCA}.

*To achieve indistinguishability (*IND-CCA*) in StE or CtEaS, the underlying encryption must be, in case the considered reduction is key-preserving, at least* IND-PCA *secure. The restriction on the reduction can be lifted if the encryption enjoys non-malleability of the key generator.* □

Theorem 7.2 (Security of EtS) *Signcryptions from EtS are* EUF-CMA *secure if the underlying signature is* EUF-CMA *secure. Moreover, they are* IND-CCA *secure if they use* SEUF-CMA *secure signature and* IND-CPA *secure encryption.* □

7.1.2 Deficiencies of the New StE and CtEtS Paradigms

Signcryptions from StE or CtEaS suffer the strong forgeability: given a signcryption on some message, one can create another signcryption on the same message without the sender's help. To circumvent this problem, we can apply the same techniques used previously for confirmer signatures, namely bind the digital signature to its corresponding signcryption. This translates for CtEaS in producing the digital signature on both the commitment and the encryption. Similarly to confirmer signatures, the new CtEaS looses the parallelism of the original one, i.e. encryption and signature can no longer be carried out in parallel, however it has the advantage of resting on cheap encryption compared to the early one. The new StE uses similarly an encryption scheme from the hybrid encryption paradigm, and the digital signature is produced on both the encapsulation of the key (used later for encryption) and the message.

Unfortunately, verifiability turns out to be a hurdle in both StE and CtEaS; the new (and old) StE paradigm encrypts the message to be signcrypted concatenated with a digital signature. As we are interested in proving the validity of the produced signcryption, we need to exploit the homomorphic properties of the signature and of the encryption schemes in order to provide proofs of knowledge of the encrypted signature and message. As a consequence, the used encryption and signature need to operate on elements from a set with a known algebraic structure rather than on bit-strings (concatenation breaks the homomorphic properties of the schemes, if any). The same remark applies to the new (and old) CtEaS paradigm as encryption is performed on the concatenation of the message to be signcrypted and the opening value of the commitment scheme.

This leaves us with only the EtS paradigm to get efficient verifiable signcryption. In fact, the sender needs simply to prove knowledge of the decryption of a given ciphertext. Also, the receiver has to prove that a message is/isn't the decryption of a given ciphertext. Such proofs are easy to carry out if one considers the already mentioned class $\mathbb{E}$. Moreover, `sigExtract` (similarly to conversion in confirmer signatures) can be made efficient for many encryption schemes from the class $\mathbb{E}$. Unfortunately, signcryptions from EtS are not anonymous, i.e. disclose the identity of the sender (anyone can check the validity of the digital signature on the ciphertext w.r.t. the sender's public key).

To sum-up, EtS provides efficient verifiability but at the expense of the sender's anonymity. StE achieves better privacy but at the expense of verifiability. It would be nice to have a technique that combines the merits of both paradigms while avoiding their drawbacks. This is the main contribution of the next section.

7.2 EtStE: A New Paradigm for Efficient Verifiable Signcryption

The core idea consists in first encrypting the message to be signcrypted using a public-key encryption scheme, then applying the new StE to the produced encryption. The result of this operation in addition to the encrypted message form the new signcryption of the message in question. In other terms, this technique can be seen as a merge between EtS and StE; thus we can term it the "Encrypt_then_Sign_then_Encrypt" paradigm (EtStE).

setup(1^κ)	: $\{\Sigma,\Gamma,(\mathcal{K},\mathcal{D})\}$.setup($1^\kappa$) ; crs $\leftarrow$ TA.setup(1^κ)
keygen$_S$(1^κ)	: Σ.keygen(1^κ)
keygen$_R$(1^κ)	: $\{\Gamma,(\mathcal{K},\mathcal{D})\}$.keygen($1^\kappa$)
signcrypt(m)	: $e_1 \leftarrow \Gamma.\mathtt{encrypt}_{\{\Gamma.pk,coins_1\}}(m)$; $(k,c) \leftarrow \mathcal{K}.\mathtt{encap}_{\{\mathcal{K}.pk,coins_2\}}()$ $(s,r) \leftarrow \Sigma.\mathtt{sign}(c\Vert e_1)$; $e_2 \leftarrow \mathcal{D}.\mathtt{encrypt}_k(s)$ return (e_1,c,e_2,r)
proveValidity(e_1,c,e_2,r)	: ZKP$\{(m,coins_1,s,coins_2)$: $e_1 = \Gamma.\mathtt{encrypt}_{\{\Gamma.pk,coins_1\}}(m)$ $\wedge$ $(c,e_2) = (\mathcal{K},\mathcal{D}).\mathtt{encrypt}_{\{\mathcal{K}.pk,coins_2\}}(s) \wedge$ $\Sigma.\mathtt{verify}_{\Sigma.pk}([s,r],c\Vert e_1) = 1\}$
unsigncrypt(e_1,c,e_2,r)	: $s \leftarrow (\mathcal{K},\mathcal{D}).\mathtt{decrypt}_{\mathcal{K}.sk}(c,e_2)$; $b \leftarrow \Sigma.\mathtt{verify}_{\Sigma.pk}([s,r],c\Vert e_1)$ if $b=0$ return $(\perp)$ else return $(\Gamma.\mathtt{decrypt}_{\Gamma.sk}(e_1))$
confirm($[e_1,c,e_2,r],m$)	: ZKP$\{(s,\Gamma.sk,\mathcal{K}.sk)$: $m = \Gamma.\mathtt{decrypt}_{\Gamma.sk}(e_1)$ $\wedge$ $s = (\mathcal{K},\mathcal{D}).\mathtt{decrypt}_{\{\mathcal{K}.sk\}}(c,e_2)$ $\wedge$ $\Sigma.\mathtt{verify}_{\Sigma.pk}([s,r],c\Vert e_1) = 1\}$
deny($[e_1,c,e_2,r],m$)	: ZKP$\{\Gamma.sk$: $m \neq \Gamma.\mathtt{decrypt}_{\Gamma.sk}(e_1)\}$ $\vee$ ZKP$\{(s,\mathcal{K}.sk)$: $s = (\mathcal{K},\mathcal{D}).\mathtt{decrypt}_{\{\mathcal{K}.sk\}}(c,e_2)$ $\wedge$ $\Sigma.\mathtt{verify}_{\Sigma.pk}([s,r],c\Vert e_1) \neq 1\}$
sigExtract($[e_1,c,e_2,r],m$)	: $s \leftarrow (\mathcal{K},\mathcal{D}).\mathtt{decrypt}_{\mathcal{K}.sk}(c,e_2)$; $b \leftarrow \Sigma.\mathtt{verify}_{\Sigma.pk}([s,r],c\Vert e_1)$ if $b=0$ return $(\perp)$ else $\{\pi \leftarrow$ NIZK$\{m = \Gamma.\mathtt{decrypt}_{\Gamma.sk}(e_1)\}$; return $(\pi,e_1,c,s,r)\}$
sigVerify($[\pi,e_1,c,s,r],m$)	: NIZK.verify(crs,π) ; $\Sigma.\mathtt{verify}_{\Sigma.pk}([s,r],c\Vert e_1)$

Fig. 7.1 The EtStE paradigm for signcryption

7.2.1 The Construction

Consider a signature scheme Σ, an encryption scheme Γ, and another encryption $(\mathcal{K},\mathcal{D})$ derived from the KEM/DEM paradigm. On input the security parameter κ, generate the parameters *param* of these schemes. Note that a trusted authority is needed to generate the common reference string for the NIZK proofs. We assume that signatures issued with Σ can be written, in a reversible way, as (s, r), where s represents the "significant" part of the signature, and r reveals no information about the signed message nor about the public signing key. Signcryptions from EtStE are described in Fig. 7.1.

7.2.2 Security Analysis

Signcryptions from EtStE meet the following strong indistinguishability notion, which captures both the anonymity of the sender and the indistinguishability of the signcryptions. The notion informally denotes the difficulty to distinguish signcryptions on an adversarially chosen message from random elements in the signcryption space.

Definition 7.1 (Strong Indistinguishability for Signcryption (SIND-CCA)**)** Let SC be a signcryption scheme, and let $\mathcal{A}$ be a PPTM. We consider the random experiment for security parameter κ and $b \xleftarrow{R} \{0,1\}$

Experiment $\mathbf{Exp}^{\mathsf{SIND\text{-}CCA}\text{-}b}_{SC,\mathcal{A}}(1^\kappa)$

1. $param \leftarrow SC.\mathtt{setup}(1^\kappa)$
2. $(sk_{\mathsf{S}}, pk_{\mathsf{S}}) \leftarrow SC.\mathtt{keygen}_{\mathsf{S}}(1^\kappa, param)$
3. $(sk_{\mathsf{R}}, pk_{\mathsf{R}}) \leftarrow SC.\mathtt{keygen}_{\mathsf{R}}(1^\kappa, param)$
4. $(m^\star, \mathcal{I}) \leftarrow \mathcal{A}^{\mathfrak{S},\mathfrak{V},\mathfrak{U},\mathfrak{C}}(pk_{\mathsf{S}}, pk_{\mathsf{R}})$
   - $\mathfrak{S} : m \longmapsto SC.\mathtt{signcrypt}_{sk_{\mathsf{S}}}(m, pk_{\mathsf{S}}, pk_{\mathsf{R}})$
   - $\mathfrak{V} : \mu \longmapsto SC.\mathtt{proveValidity}_{coins}(\mu, pk_{\mathsf{S}}, pk_{\mathsf{R}})$
   - $\mathfrak{U} : \mu \longmapsto SC.\mathtt{unsigncrypt}_{sk_{\mathsf{R}}}(\mu, pk_{\mathsf{S}}, pk_{\mathsf{R}})$
   - $\mathfrak{C} : (\mu, m) \longmapsto SC.\{\mathtt{confirm}, \mathtt{deny}\}_{sk_{\mathsf{R}}}(\mu, m, pk_{\mathsf{S}}, pk_{\mathsf{R}})$
   - $\mathfrak{P} : (\mu, m) \longmapsto SC.\mathtt{sigExtract}_{sk_{\mathsf{R}}}(\mu, m, pk_{\mathsf{S}}, pk_{\mathsf{R}})$
5. $\mu_1^\star \leftarrow SC.\mathtt{signcrypt}_{sk_{\mathsf{S}}}(m^\star, pk_{\mathsf{S}}, pk_{\mathsf{R}})$; $\mu_0^\star \xleftarrow{R} SC.\mathsf{space}$
6. $d \leftarrow \mathcal{A}^{\mathfrak{S},\mathfrak{V},\mathfrak{U},\mathfrak{C}}(\mathcal{I}, \mu_b^\star, pk_{\mathsf{S}}, pk_{\mathsf{C}})$
   - $\mathfrak{S} : m \longmapsto SC.\mathtt{signcrypt}_{sk_{\mathsf{S}}}(m, pk_{\mathsf{S}}, pk_{\mathsf{R}})$
   - $\mathfrak{V} : \mu(\neq \mu^\star) \longmapsto SC.\mathtt{proveValidity}_{coins}(\mu, pk_{\mathsf{S}}, pk_{\mathsf{R}})$
   - $\mathfrak{U} : \mu(\neq \mu^\star) \longmapsto SC.\mathtt{unsigncrypt}_{sk_{\mathsf{R}}}(\mu, pk_{\mathsf{S}}, pk_{\mathsf{R}})$
   - $\mathfrak{C} : (\mu, m)(\neq (\mu^\star, m^\star)) \longmapsto SC.\{\mathtt{confirm}, \mathtt{deny}\}_{sk_{\mathsf{R}}}(\mu, m, pk_{\mathsf{S}}, pk_{\mathsf{R}})$
   - $\mathfrak{P} : (\mu, m)(\neq (\mu^\star, m^\star)) \longmapsto SC.\mathtt{sigExtract}_{sk_{\mathsf{R}}}(\mu, m, pk_{\mathsf{S}}, pk_{\mathsf{R}})$
7. return (d)

We define the *advantage* of $\mathcal{A}$ *via*:

$$\mathbf{Adv}^{\mathsf{SIND\text{-}CCA}}_{SC,\mathcal{A}}(1^\kappa) = \left| \Pr\left[\mathbf{Exp}^{\mathsf{SIND\text{-}CCA}\text{-}b}_{SC,\mathcal{A}}(1^\kappa) = b\right] - \frac{1}{2} \right|.$$

Given $(t, q_s, q_v, q_u, q_{cd}, q_e) \in \mathbb{N}^6$ and $\varepsilon \in [0,1]$, $\mathcal{A}$ is called a $(t, \varepsilon, q_s, q_v, q_u, q_{cd}, q_e)$-SIND-CCA adversary against SC if, running in time t and issuing q_s queries to the `signcrypt` oracle, q_v queries to the `proveValidity` oracle, q_u queries to the `unsigncrypt` oracle, q_{cd} queries to the $\{\mathtt{confirm}, \mathtt{deny}\}$ oracle, and q_e to the `sigExtract` oracle, $\mathcal{A}$ has $\mathbf{Adv}^{\mathsf{SIND\text{-}CCA}}_{SC,\mathcal{A}}(1^\kappa) \geq \varepsilon$. The scheme SC is $(t, \varepsilon, q_s, q_v, q_u, q_{cd}, q_e)$-SIND-CCA secure if no $(t, \varepsilon, q_s, q_v, q_u, q_{cd}, q_e)$-SIND-CCA adversary against it exists.

Theorem 7.3 *Given* $(t, q_s) \in \mathbb{N}^2$ *and* $\epsilon \in [0,1]$, *the above construction is* (t, ϵ, q_s)-EUF-CMA *secure if the underlying digital signature scheme is* (t, ϵ, q_s)-EUF-CMA *secure.* □

Theorem 7.4 *Given* $(t, q_s, q_v, q_u, q_{cd}, q_e) \in \mathbb{N}^6$ *and* $(\epsilon, \epsilon', \epsilon_s, \epsilon_e, \epsilon_d) \in [0,1]^5$, *the construction proposed above is* $(t, \epsilon, q_s, q_v, q_u, q_{cd}, q_e)$-SIND-CCA *secure if it uses a* (t, ϵ_s, q_s)-SEUF-CMA *secure digital signature, a* (t, ϵ_e)-INV-CPA *secure encryption, a* (t, ϵ_d)-INV-OT *secure DEM with injective encryption, and a* $(t + q_s(q_v + q_u + q_{cd} + q_e), \epsilon(1-\epsilon_e)(1-\epsilon_d)(1-\epsilon_s)^{q_v+q_{cd}+q_u+q_{pv}})$-IND-CPA *secure KEM.*

Proof (Sketch) From an SIND-CCA adversary $\mathcal{A}$ against the construction, we construct an algorithm $\mathcal{R}$ that IND-CPA break the KEM underlying the construction. $\mathcal{R}$ gets the public parameters of the KEM from her challenger and chooses further the remaining building blocks, namely, the DEM, the signature, and the encryption scheme. Simulation of $\mathcal{A}$'s environment is done using the key pairs of the used signature and encryption schemes, in addition to a list in which $\mathcal{R}$ maintains the queries, their responses and the intermediate values used to generate these responses.

Eventually, $\mathcal{A}$ outputs a challenge message m. $\mathcal{R}$ will encrypt, in e, the message m. Next, she produces a signature (s, r) on $c\|e$, where (c, k) is her challenge. Finally, $\mathcal{R}$ encrypts s in $e_{\mathcal{D}}$ using k, and outputs $\mu = (e, c, e_{\mathcal{D}}, r)$ as a challenge signcryption. Since the used encryption is INV-CPA secure by assumption, then information about m can only leak from $(c, e_{\mathcal{D}}, r)$. If k is the decapsulation of c, then μ is a valid signcryption of m, otherwise it is a random element from the signcryption space due to the assumptions on the used components (encryption scheme is INV-CPA, the DEM is INV-OT, and finally r reveals no information about e nor about sender's key). The rest follows as in the proof of Theorem 4.2. □

7.2.3 *Practical Instantiations*

The `proveValidity` and {`confirm`, `deny`} protocols comprise the following sub-protocols:

1. Proving knowledge of the decryption of a ciphertext produced using the encryption scheme Γ.
2. Proving that a message is/isn't the decryption of some ciphertext produced using Γ.
3. Proving knowledge of the decryption of a ciphertext produced using $(\mathcal{K}, \mathcal{D})$, and that this decryption forms a valid/invalid digital signature issued using Σ on some known string.

It is natural to instantiate the encryption Γ from class $\mathbb{E}$ described in Definition 4.3. With this choice, the first sub-protocol can be efficiently carried out as depicted in Fig. 4.3, whereas the second sub-protocol can be implemented as explained in Sect. 6.2.2. Moreover, one can consider encryptions from class $\mathbb{E}$ that are derived from the KEM/DEM paradigm, in addition to signatures from class $\mathbb{S}$ described in Definition 4.2; with this choice, the last sub-protocol boils down to the protocol depicted in Fig. 4.4.

Finally, for the `sigExtract` algorithm, we refer to the solutions adopted in confirmer signatures (described in Sect. 6.2.3) when it comes to producing a NIZK proof of the correctness of a decryption.

References

An JH, Dodis Y, Rabin T (2002) On the security of joint signature and encryption. In: Knudsen LR (ed) Advances in cryptology - EUROCRYPT 2002. LNCS, vol 2332. Springer, Heidelberg, pp 83–107

Blum M, Feldman P, Micali S (1988) Non-interactive zero-knowledge and its applications (extended abstract). In: Simon J (ed) STOC. ACM Press, New York, pp 103–112

Chiba D, Matsuda T, Schuldt JN, Matsuura K (2011) Efficient generic constructions of signcryption with insider security in the multi-user setting. In: Lopez J, Tsudik G (eds) Applied cryptography and network security. LNCS, vol 6715. Springer, Heidelberg, pp 220–237

Goldreich O, Micali S, Wigderson A (1986) How to prove all NP-statements in zero-knowledge, and a methodology of cryptographic protocol design. In: Odlyzko AM (ed) CRYPTO. LNCS, vol 263. Springer, Heidelberg, pp 171–185

Matsuda T, Matsuura K, Schuldt J (2009) Efficient constructions of signcryption schemes and signcryption composability. In: Roy B, Sendrier N (eds) IndoCrypt, vol 5922. Springer, Berlin/Heidelberg, pp 321–342

Chapter 8
Multi-User Security

Abstract Hitherto, we have considered only a network of two users: signer/confirmer in case of confirmer signatures, and sender/receiver in the signcryption case. This setting is too simplistic to represent reality, where it is customary to have a network of many users that want to exchange signcrypted messages. Also, it is not uncommon in case of confirmer signatures, to have many signers that share the same confirmer, or conversely a signer who has many confirmers. We tackle in this chapter the issue of multi-user security; we first describe the concerns that arise in this extended model, then we formalize these issues in new security definitions, and finally, we give the new analogs of StE, CtEtS, and EtS in the multi-user setting.

8.1 Motivation and Definition

A primitive secure in the two-user setting does not necessarily mean that it preserves this security in the multi-user setting. In fact, the unforgeability adversary in the latter mode is allowed to return a forgery on a message $m^\star$ that may have been queried before but w.r.t. a confirmer/receiver's key different from the target key $pk^\star$. Moreover, the invisibility/indistinguishability adversary is allowed to ask the conversion/unsigncryption of the challenge w.r.t. any confirmer/receiver's key except that of the target confirmer/receiver.

Therefore, if we move on to the multi-user setting, several new concerns arise.

First, users must now have identities. We do not impose any constraints on the identities, except that they should be easily recognizable by everyone in the network, and that users can easily obtain the public key of a given identity.

Next, we should change the syntax of the protocols/algorithms underlying the studied primitives so as to include the identities or public keys of the playing users. Therefore, if pk_{S} and pk_{C} (pk_{R}) are the public keys of the involved signer and confirmer (receiver) respectively, then they must be part of the input and output to all algorithms/protocols underlying the confirmer signature (signcryption) scheme.

Moreover, the adversary, in the multi-user setting, may collude with other users in order to break a given security notion. It is then natural to grant him access, in addition to the oracles he had access to in the two-user setting, to all private keys except those of the users involved in the challenge experiment. Besides, to break unforgeability, the adversary has to come up with a valid confirmer signature

L. El Aimani, *Verifiable Composition of Signature and Encryption*,
https://doi.org/10.1007/978-3-319-68112-2_8

(signcryption) on some message m that was not queried for signature (signcryption) with respect to the challenge keys pk_{S} and pk_{C} (pk_{R}) of the signer and the confirmer (receiver) resp. Similarly, to break invisibility (indistinguishability) the adversary has to come up with messages m_0 and m_1 such that he cannot distinguish their corresponding confirmer signatures (signcryptions). Of course, given a challenge confirmer signature (signcryption) μ w.r.t. identities pk_{S} and pk_{C} (pk_{R}), the adversary is disallowed to ask its conversion/verification (unsigncryption/verification) w.r.t. these challenge identities.

8.1.1 Formal Security Model

We define in this subsection unforgeability and invisibility of confirmer signatures in the multi-user setting. Multi-user unforgeability and indistinguishability for signcryption are similar and will be omitted.

In both definitions, S and C denote respectively the sets of signers' and of confirmers' identities. Both sets are finite and have respective cardinalities N and M.

Definition 8.1 (Multi-User Unforgeability for Confirmer Signatures) Let CS be a CDCS scheme and $\mathcal{A}$ be a PPTM. We consider the following experiment where κ is a security parameter.

Experiment $\mathbf{Exp}_{CS,\mathcal{A}}^{\mathsf{EUF\text{-}CMA\text{-}multi}}(1^\kappa)$

1. $param \leftarrow \mathtt{setup}(1^\kappa)$
2. $S \leftarrow \{\mathsf{S}_1, \ldots, \mathsf{S}_\mathsf{N}\}$; $C \leftarrow \{\mathsf{C}_1, \ldots, \mathsf{C}_\mathsf{M}\}$
3. $(i^\star, j^\star) \leftarrow \mathcal{A}(param, S, C)$
4. $(pk_{\mathsf{S}}^{i^\star}, sk_{\mathsf{S}}^{i^\star}) \leftarrow CS.\mathtt{keygen}_{\mathsf{S}_{i^\star}}(1^\kappa)$
5. $(pk_{\mathsf{C}}^{j^\star}, sk_{\mathsf{C}}^{j^\star}) \leftarrow \mathcal{A}(pk_{\mathsf{S}}^{i^\star})$
6. $(pk_{\mathsf{S}}^{\star}, sk_{\mathsf{S}}^{\star}) \leftarrow (pk_{\mathsf{S}}^{i^\star}, sk_{\mathsf{S}}^{i^\star})$; $(pk_{\mathsf{C}}^{\star}, sk_{\mathsf{C}}^{\star}) \leftarrow (pk_{\mathsf{C}}^{j^\star}, sk_{\mathsf{C}}^{j^\star})$
7. $(m^\star, \mu^\star) \leftarrow \mathcal{A}^{\mathfrak{S}}(pk_{\mathsf{S}}^{\star}, sk_{\mathsf{C}}^{\star}, pk_{\mathsf{C}}^{\star})$
   $\mathfrak{S} : (m, pk_{\mathsf{S}}^{\star}, pk_{\mathsf{C}}) \longmapsto CS.\mathtt{sign}_{sk_{\mathsf{S}}^{\star}}(m, pk_{\mathsf{C}})$
8. return 1 if and only if:
   - $\mathtt{verify}(\mu^\star, m^\star, pk_{\mathsf{S}}^{\star}, pk_{\mathsf{C}}^{\star}) = 1$
   - $(m^\star, pk_{\mathsf{S}}^{\star}, pk_{\mathsf{C}}^{\star})$ was not queried to $\mathfrak{S}$

We define the *advantage* of $\mathcal{A}$ *via*:

$$\mathbf{Adv}_{CS,\mathcal{A}}^{\mathsf{EUF\text{-}CMA\text{-}multi}}(1^\kappa) = \Pr\left[\mathbf{Exp}_{CS,\mathcal{A}}^{\mathsf{EUF\text{-}CMA\text{-}multi}}(1^\kappa) = 1\right].$$

CS is (t, ϵ, q_s)-EUF-CMA-multi secure if there exists no adversary operating in time t and issuing q_s queries to the signing oracle, that wins the game defined in the experiment above with advantage greater that ϵ. The probability is taken over all the coin tosses.

Definition 8.2 (Multi-User Invisibility for Confirmer Signatures) Let CS be a CDCS scheme and $\mathcal{A}$ be a PPTM. We consider the following experiment where κ is a security parameter.

Experiment $\mathbf{Exp}_{CS,\mathcal{A}}^{\mathsf{INV\text{-}CMA\text{-}multi}}(1^\kappa)$

1. $param \leftarrow \mathtt{setup}(1^\kappa)$
2. $S \leftarrow \{\mathsf{S}_1, \ldots, \mathsf{S}_\mathsf{N}\}; C \leftarrow \{\mathsf{C}_1, \ldots, \mathsf{C}_\mathsf{M}\}$
3. $(i^\star, j^\star) \leftarrow \mathcal{A}(param, S, C)$
4. $(pk_\mathsf{S}^{i^\star}, sk_\mathsf{S}^{i^\star}) \leftarrow CS.\mathtt{keygen}_{\mathsf{S}_{i^\star}}(1^\kappa)$
5. $(pk_\mathsf{C}^{j^\star}, sk_\mathsf{C}^{j^\star}) \leftarrow CS.\mathtt{keygen}_{\mathsf{C}_{j^\star}}(1^\kappa)$
6. $(pk_\mathsf{S}^\star, sk_\mathsf{S}^\star) \leftarrow (pk_\mathsf{S}^{i^\star}, sk_\mathsf{S}^{i^\star})$; $(pk_\mathsf{C}^\star, sk_\mathsf{C}^\star) \leftarrow (pk_\mathsf{C}^{j^\star}, sk_\mathsf{C}^{j^\star})$
7. $(m_0^\star, m_1^\star, \mathcal{I}) \leftarrow \mathcal{A}^{\mathfrak{S},\mathfrak{Cv},\mathfrak{V}}(param, pk_\mathsf{S}^\star, pk_\mathsf{C}^\star)$
   - $\mathfrak{S} : (m, pk_\mathsf{S}^\star, pk_\mathsf{C}^\star) \longmapsto CS.\mathtt{sign}_{sk_\mathsf{S}^\star}(m, pk_\mathsf{C}^\star)$
   - $\mathfrak{Cv} : (\mu, m, pk_\mathsf{S}, pk_\mathsf{C}^\star) \longmapsto CS.\mathtt{convert}_{sk_\mathsf{C}^\star}(\mu, m, pk_\mathsf{S}, pk_\mathsf{C}^\star)$
   - $\mathfrak{V} : (\mu, m, pk_\mathsf{S}, pk_\mathsf{C}^\star) \longmapsto CS.\{(\mathsf{s})\mathtt{confirm}, \mathtt{deny}\}_{\{coins_\mu \vee sk_\mathsf{C}^\star\}}(\mu, m)$
8. $b \overset{R}{\leftarrow} \{0,1\}$; $\mu^\star \leftarrow CS.\mathtt{sign}_{sk_\mathsf{S}^\star}(m_b^\star, pk_\mathsf{C}^\star)$
9. $b^\star \leftarrow \mathcal{A}^{\mathfrak{S},\mathfrak{Cv},\mathfrak{V}}(\text{guess}, \mathcal{I}, \mu^\star, pk_\mathsf{S}^\star, pk_\mathsf{C}^\star)$
   - $\mathfrak{S} : (m, pk_\mathsf{S}^\star, pk_\mathsf{C}^\star) \longmapsto CS.\mathtt{sign}_{sk_\mathsf{S}^\star}(m, pk_\mathsf{C}^\star)$
   - $\mathfrak{Cv} : (\mu, m, pk_\mathsf{S}, pk_\mathsf{C}^\star)(\neq (\mu^\star, m_i^\star, pk_\mathsf{S}^\star, pk_\mathsf{C}^\star), i = 0,1) \longmapsto CS.\mathtt{convert}_{sk_\mathsf{C}^\star}(\mu, m, pk_\mathsf{S}, pk_\mathsf{C}^\star)$
   - $\mathfrak{V} : (\mu, m, pk_\mathsf{S}, pk_\mathsf{C}^\star)(\neq (\mu^\star, m_i^\star, pk_\mathsf{S}^\star, pk_\mathsf{C}^\star), i = 0,1) \longmapsto CS.\{(\mathsf{s})\mathtt{confirm}, \mathtt{deny}\}(\mu, m)$
10. return $(b = b^\star)$

We define the *advantage* of $\mathcal{A}$ *via*:

$$\mathbf{Adv}_{CS,\mathcal{A}}^{\mathsf{INV\text{-}CMA\text{-}multi}}(1^\kappa) = \left| \Pr\left[\mathbf{Exp}_{CS,\mathcal{A}}^{\mathsf{INV\text{-}CMA\text{-}multi}}(1^\kappa) = 1\right] - \frac{1}{2} \right|.$$

CS is $(t, \epsilon, q_s, q_v, q_{sc})$-INV-CMA-multi secure if no adversary operating in time t, issuing q_s queries to the signing oracle (followed potentially by queries to the `sconfirm` oracle), q_v queries to the confirmation/denial oracles and q_{sc} queries to the selective conversion oracle that wins the game defined above with advantage greater that ϵ. The probability is taken over all the coin tosses.

Remark 8.1 In the notions above, the unforgeability adversary is allowed to ask signatures with respect to any confirmer's key pk_C and not necessarily the challenge key $pk_\mathsf{C}^\star$. Similarly, the invisibility adversary is allowed to ask conversion/verification queries with respect to any signer's key pk_S and not just $pk_\mathsf{S}^\star$.

8.1.2 Extension to Multi-User Security

Many works have proposed simple tweaks to turn a confirmer signature (signcryption) scheme secure in the two-user setting into a full-fledged multi-user secure one. For instance the authors in An et al. (2002) propose the following simple changes to upgrade security to the multi-user setting:

1. Whenever *encrypting* something, include the identity (e.g. public key) of the *signer* together with the encrypted message.
2. Whenever *signing* something, include in addition to the message to be signed, the identity of the confirmer, in case of confirmer signatures, or of the receiver in case of signcryption.
3. On the receiving side, whenever the identities of the protagonists do not match what is expected, output $\perp$.

While the changes relative to signing can apply directly to our constructions, the proposed alternate encryption seems to pose a problem for verifiability. In fact, the verification protocols underlying our primitives often include a proof of knowledge (with/out interaction with the verifier) of a message underlying a given ciphertext. We showed how to efficiently produce such proofs if the message is a group element and the encryption algorithm satisfies some further properties. Therefore, encrypting the concatenation of the message and the user identity destroys the algebraic structure of the message space and hinders as a consequence the required proofs of knowledge.

The works (Chiba et al. 2011; Matsuda et al. 2009) propose several optimizations to the well known paradigms StE, CtEaS, and EtS, in the context of signcryption, that achieve multi-user insider security. The key tool underlying their constructions is tag-based encryption. However, as verifiability is not considered, the proposed constructions do not fit into our framework where efficient proofs of knowledge of a decryption are a necessary brick in the verification procedures.

In the rest of this chapter, we build from An et al. (2002), Matsuda et al. (2009), and Chiba et al. (2011) and extend our constructions so as to provide multi-user security without compromising their verifiability.

8.2 New Paradigms

We are now about to describe the new analogs of the classical paradigms in the multi-user setting. The main ideas underlying these constructions are:

1. Whenever *encrypting* something, use a tag-based encryption scheme to encrypt the message where the tag is set to the public key of the *signer*.
2. Whenever *signing* something, include in addition to the message to be signed, the public key of the confirmer in case of confirmer signatures, or the receiver's public key in case of signcryption.

As seen before, we need the following building blocks: a signature scheme Σ, an encryption scheme Γ, an encryption scheme $(\mathcal{K}, \mathcal{D})$ from the KEM/DEM paradigm, and finally a commitment scheme Ω. The only difference lies in the fact that both Γ and $\mathcal{K}$ support tags.

Recall that in the StE paradigm, signatures generated using the signature scheme Σ can be efficiently transformed in a reversible way to a pair (s, r) where s is the "useful" part of the signature and r reveals no information about the signed message nor about ($\Sigma.sk$, $\Sigma.pk$).

Finally, it is needless to say that verification or conversion of confirmer signatures from CtEtS (EtS) are carried out only if the underlying digital signature is valid w.r.t. the given the commitment (encryption).

8.2.1 Security Analysis

For unforgeability, whatever security was proven in the two-user setting remains unchanged in the multi-user case. However, the tag-based encryption underlying the aforementioned paradigms need to be IND-st-wCCA secure in order to achieve outsider invisibility.

We sketch in this paragraph the security analysis of the multi-user StE; analyses of the other paradigms are similar and thus will be omitted.

Theorem 8.1 (Multi-User StE) *The multi-user StE, described in Fig. 8.1, is* EUF-CMA-multi *secure if the underlying signature scheme is* EUF-CMA *secure. Moreover, it is* INV-CMA-multi *secure if it uses a* SEUF-CMA *signature, an* IND-st-wCCA *secure tag-based KEM, and an* INV-OT *secure DEM with injective encryption.*

$\mathcal{CS}.\mathtt{setup}(1^\kappa)$	: $\Sigma.\mathtt{setup}(1^\kappa)$; $\mathcal{K}.\mathtt{setup}(1^\kappa)$; $\mathcal{D}.\mathtt{setup}(1^\kappa)$
$\mathcal{CS}.\mathtt{keygen}_S(1^\kappa)$	: $\Sigma.\mathtt{keygen}(1^\kappa)$
$\mathcal{CS}.\mathtt{keygen}_C(1^\kappa)$	: $\mathcal{K}.\mathtt{keygen}(1^\kappa)$
$\mathcal{CS}.\mathtt{sign}(m,\Sigma.pk,\mathcal{K}.pk)$	: $(c,k) \leftarrow \mathcal{K}.\mathtt{encap}_{\{\mathcal{K}.pk,coins\}}(\Sigma.pk)$
	$(s,r) \leftarrow \Sigma.\mathtt{sign}_{\Sigma.sk}(c\|m\|\mathcal{K}.pk)$; $e \leftarrow \mathcal{D}.\mathtt{encrypt}_k(s)$
	return (c,e,r)
$\mathcal{CS}.\mathtt{sconfirm}([c,e,r],m,\Sigma.pk,\mathcal{K}.pk)$	: ZKP$\{(s,coins): (c,e) = (\mathcal{K},\mathcal{D}).\mathtt{encrypt}_{\{\mathcal{K}.pk,coins\}}(s,\Sigma.pk)\ \wedge$
	$\Sigma.\mathtt{verify}_{\Sigma.pk}([s,r],c\|m\|\mathcal{K}.pk) = 1\}$
$\mathcal{CS}.\mathtt{confirm}([c,e,r],m,\Sigma.pk,\mathcal{K}.pk)$	: ZKP$\{(s,\mathcal{K}.sk): s = (\mathcal{K},\mathcal{D}).\mathtt{decrypt}_{\{\mathcal{K}.sk\}}([c,e],\Sigma.pk)\ \wedge$
	$\Sigma.\mathtt{verify}_{\Sigma.pk}([s,r],c\|m\|\mathcal{K}.pk) = 1\}$
$\mathcal{CS}.\mathtt{deny}([c,e,r],m,\Sigma.pk,\mathcal{K}.pk)$	: ZKP$\{(s,\mathcal{K}.sk): s = (\mathcal{K},\mathcal{D}).\mathtt{decrypt}_{\{\mathcal{K}.sk\}}([c,e],\Sigma.pk)\ \wedge$
	$\Sigma.\mathtt{verify}_{\Sigma.pk}([s,r],c\|m\|\mathcal{K}.pk) = 0\}$
$\mathcal{CS}.\mathtt{convert}([c,e,r],m,\Sigma.pk,\mathcal{K}.pk)$	: $s \leftarrow (\mathcal{K},\mathcal{D}).\mathtt{decrypt}_{\mathcal{K}.sk}([c,e],\Sigma.pk)$
	if $\Sigma.\mathtt{verify}_{\Sigma.pk}([s,r],c\|m\|\mathcal{K}.pk) = 0$ return $(\perp)$
	else return (c,s,r)

Fig. 8.1 Multi-user StE for confirmer signatures

We first note that, unlike the two-user setting, confirmer signatures from the multi-user StE are not indistinguishable from random elements in the signature space since they include the identities of the signer and of the confirmer.

Proof Unforgeability is straightforward.

For invisibility, we focus only on the differences with the two-user setting and avoid the redundancies.

After the adversary specifies the challenge identities (of the signer and of the confirmer), the reduction chooses a suitable signature scheme, generates the signer's key pair $(sk_{\mathsf{S}}^{\star}, pk_{\mathsf{S}}^{\star})$ and hands $pk_{\mathsf{S}}^{\star}$ to her challenger as a challenge tag. She gets is response $pk_{\mathsf{C}}^{\star}$ as a public key of the tag-based encryption subject of the attack. The reduction forwards then $(pk_{\mathsf{S}}^{\star}, pk_{\mathsf{C}}^{\star})$ to the adversary in addition to further public parameters relative to the used building blocks.

We distinguish two types of queries: those with respect to $pk_{\mathsf{S}}^{\star}$ and those with respect to a different signer's key.

For the first set of queries, the reduction proceeds exactly as in the two-user case, i.e. use a list of records that maintains the queried messages, the output signatures and the intermediate values used to generate these signatures. In fact, as shown previously, such a list turns out to be sufficient to handle the verification/conversion queries provided the signature scheme is strongly unforgeable.

For the latter type of queries, the reduction proceeds as dictated by the standard algorithms/protocols. In fact, she can query her challenger for the decryption of any ciphertext w.r.t. any tag different from the challenge tag $pk_{\mathsf{S}}^{\star}$.

Once the adversary hands the challenge messages m_0 and m_1. The reduction chooses at random a bit $b \overset{R}{\leftarrow} \{0,1\}$ and uses her challenge $(c^{\star}, k^{\star})$ to compute a digital signature $(s^{\star}, r^{\star})$ on $c^{\star}\|m_b\|pk_{\mathsf{C}}^{\star}$. She further encrypts $s^{\star}$ in $e^{\star}$ using the DEM encryption algorithm and $k^{\star}$. Finally, she hands $\mu^{\star} = (c^{\star}, e^{\star}, r^{\star})$ to the adversary as a challenge confirmer signature. Note that if $c^{\star}$ is the encapsulation of the key $k^{\star}$, then $\mu^{\star}$ is a valid confirmer signature on m_b, otherwise it is not a valid signature on either message. In fact, $(c^{\star}, e^{\star})$ is a random element in the ciphertext space since $k^{\star}$ is a random key and the used DEM is INV-OT secure. Moreover, the used signature scheme decrees that $r^{\star}$ is statistically independent of both the signed message and the signing key. Therefore, when the adversary outputs his guess b' of the message underlying $\mu^{\star}$, the reduction returns the bit $b = b'$ to her challenger. □

Theorem 8.2 (Multi-User CtEtS) *The multi-user CtEtS, described in Fig. 8.2, is* EUF-CMA-Multi *secure if the underlying signature scheme is* EUF-CMA *secure and the used commitment is computationally binding. Moreover, it is* INV-CMA-Multi *secure if it uses a* SEUF-CMA *secure signature, an* IND-st-wCCA *secure tag-based encryption, and an injective and computationally hiding and binding commitment.* □

Theorem 8.3 (Multi-User EtS) *The multi-user EtS, described in Fig. 8.3, is* EUF-CMA-multi *secure if the underlying signature scheme is* EUF-CMA *secure. Moreover, it is* INV-CMA-multi *secure if it uses a* SEUF-CMA *secure signature and an* IND-st-wCCA *secure tag-based encryption.* □

$CS.\mathtt{setup}(1^\kappa)$	: $\Sigma.\mathtt{setup}(1^\kappa)$; $\Gamma.\mathtt{setup}(1^\kappa)$; $\Omega.\mathtt{setup}(1^\kappa)$
$CS.\mathtt{keygen_S}(1^\kappa)$	: $\Sigma.\mathtt{keygen}(1^\kappa)$
$CS.\mathtt{keygen_C}(1^\kappa)$	: $\Gamma.\mathtt{keygen}(1^\kappa)$
$CS.\mathtt{sign}(m,\Sigma.pk,\Gamma.pk)$	: $c \leftarrow \Omega.\mathtt{commit}(m,r)$; $e \leftarrow \Gamma.\mathtt{encrypt}_{\{\Gamma.pk,coins\}}(r,\Sigma.pk)$
	$\sigma \leftarrow \Sigma.\mathtt{sign}_{\Sigma.sk}(e\|c\|\Gamma.pk)$; return (c,e,σ)
$CS.\mathtt{sconfirm}([c,e,\sigma],m,\Sigma.pk,\Gamma.pk)$	: ZKP$\{(r,coins): c = \Omega.\mathtt{commit}(m,r)\ \wedge$
	$e = \Gamma.\mathtt{encrypt}_{\{\Gamma.pk,coins\}}(r,\Sigma.pk)\}$
$CS.\mathtt{confirm}(\{c,e,\sigma\},m,\Sigma.pk,\Gamma.pk)$	: ZKP$\{(r,\Gamma.sk): c = \Omega.\mathtt{commit}(m,r)\ \wedge$
	$r = \Gamma.\mathtt{decrypt}_{\Gamma.sk}(e,\Sigma.pk)\}$
$CS.\mathtt{deny}(\{c,e,\sigma\},m,\Sigma.pk,\Gamma.pk)$	: ZKP$\{(r,\Gamma.sk): c \neq \Omega.\mathtt{commit}(m,r)\ \wedge$
	$r = \Gamma.\mathtt{decrypt}_{\Gamma.sk}(e,\Sigma.pk)\}$
$CS.\mathtt{convert}(\{c,e,\sigma\},m,\Sigma.pk,\Gamma.pk)$	: $r \leftarrow \Gamma.\mathtt{decrypt}_{\Gamma.sk}(e,\Sigma.pk)$
	if $c = \Omega.\mathtt{commit}(m,r)$ return (r,c,σ) else return $(\perp)$

Fig. 8.2 Multi-user CtEtS for confirmer signatures

$CS.\mathtt{setup}(1^\kappa)$	: $\Sigma.\mathtt{setup}(1^\kappa)$; $\Gamma.\mathtt{setup}(1^\kappa)$; crs $\leftarrow$ TA.$\mathtt{setup}(1^\kappa)$
$CS.\mathtt{keygen_S}(1^\kappa)$	: $\Sigma.\mathtt{keygen}(1^\kappa)$
$CS.\mathtt{keygen_C}(1^\kappa)$	: $\Gamma.\mathtt{keygen}(1^\kappa)$
$CS.\mathtt{sign}(m,\Sigma.pk,\Gamma.pk)$	: $c \leftarrow \Gamma.\mathtt{encrypt}_{\{\Gamma.pk,coins_c\}}(m,\Sigma.pk)$
	$\sigma \leftarrow \Sigma.\mathtt{sign}_{\Sigma.sk}(c\|\Gamma.pk)$; return $[c,\sigma]$
$CS.\mathtt{sconfirm}([c,\sigma],m,\Sigma.pk,\Gamma.pk)$	: ZKP$\{coins_c: c = \Gamma.\mathtt{encrypt}_{\{\Gamma.pk,coins_c\}}(m,\Sigma.pk)\}$
$CS.\mathtt{confirm}([c,\sigma],m,\Sigma.pk,\Gamma.pk,)$	: ZKP$\{\Gamma.sk: m = \Gamma.\mathtt{decrypt}_{\Gamma.sk}(c,\Sigma.pk)\}$
$CS.\mathtt{deny}([c,\sigma],m,\Sigma.pk,\Gamma.pk)$	: ZKP$\{\Gamma.sk: m \neq \Gamma.\mathtt{decrypt}_{\Gamma.sk}(c,\Sigma.pk)\}$
$CS.\mathtt{convert}([c,\sigma],m,\Sigma.pk,\Gamma.pk)$	: $\pi \leftarrow$ NIZK$\{m = \Gamma.\mathtt{decrypt}_{\Gamma.sk}(c,\Sigma.pk)\}$; return $[\pi,c,\sigma]$

Fig. 8.3 Multi-user EtS for confirmer signatures

Remark 8.2 (Multi-User Signcryption) The above results on multi-user security for confirmer signatures apply literally to the signcryption case. For instance, the multi-user EtS described in Fig. 8.3 for confirmer signatures can be also used for signcryption.

Note however that there will be no need to describe an analog of the EtStE paradigm in the multi-user setting. Actually, this paradigm was introduced to amend EtS that discloses the signer's key and violates consequently anonymity of signcryptions. This disclosure is now inevitable in a multi-user setting as both the signer's and receiver's identities are part of the generated signcryption.

8.2.2 Performance

We first define the following class of tag-based encryption.

Definition 8.3 (Class $\mathbb{T}$ of Homomorphic Tag-Based Encryption) $\mathbb{T}$ is the set of tag-based encryption schemes Γ that have the following properties:

1. The message space is a group $\mathbb{G}$ w.r.t. some binary operation $*$ and the ciphertext space C is a set equipped with some binary operation $\circ_e$.
2. Let $s \in \mathbb{G}$ be a message and e its encryption with respect to a given key pk and a given tag t. On the common input pk, t, s, and e, there exists an efficient zero-knowledge proof of s being the decryption of e with respect to tag t under the key pk. The private input of the prover is either the private key of the scheme or the randomness used to produce the encryption e.
3. $\forall pk\ \forall s, s' \in \mathbb{G}\ \forall t$:

$$\Gamma.\mathtt{encrypt}_{pk}(s * s', t) = \Gamma.\mathtt{encrypt}_{pk}(s, t) \circ_e \Gamma.\mathtt{encrypt}_{pk}(s', t)$$

   Moreover, given the randomness used to encrypt s in $\Gamma.\mathtt{encrypt}_{pk}(s,t)$ and s' in $\Gamma.\mathtt{encrypt}_{pk}(s',t)$, one can deduce (using only the public parameters) the randomness used to produce $\Gamma.\mathtt{encrypt}_{pk}(s,t) \circ_e \Gamma.\mathtt{encrypt}_{pk}(s',t)$ on $s * s'$.

Examples of encryption schemes in the above class include Kiltz' (2006) and Cash et al.'s (Fig. 8.4) (Cash et al. 2009) tag-based encryption schemes. In fact, the encryption algorithm in both schemes is homomorphic w.r.t. the message, and both schemes accept efficient proofs of the correctness of a decryption, namely the proof of knowledge of discrete logarithms which satisfy known predicates (linear relations).

Theorem 8.4 *Encryption schemes from class $\mathbb{T}$ accept perfect zero-knowledge proofs of knowledge of a decryption.*

Proof If we extend the protocol depicted in Fig. 4.3 to accept tags (i.e. input the tag, that is part of the common input, to the encryption and decryption algorithms), then

$\mathsf{setup}(1^\kappa)$	: Choose a group $(\mathbb{G} = \langle g \rangle, \cdot)$ with prime order d
$\mathsf{keygen}(1^\kappa)$	: $x_1, \widetilde{x}_1, x_2, \widetilde{x}_2 \xleftarrow{R} \mathbb{Z}_d$; $X_i \leftarrow g^{x_i}$ and $\widetilde{X}_i \leftarrow g^{\widetilde{x}_i}$ for $i = 1,2$ $pk \leftarrow \{X_i, \widetilde{X}_i\}_{i=1,2}$; $sk \leftarrow \{x_i, \widetilde{x}_i\}_{i=1,2}$
$\mathsf{encrypt}(m,t)$	: $r \xleftarrow{R} \mathbb{Z}_d$ $c_1 \leftarrow g^r$; $c_2 \leftarrow (X_1^t \widetilde{X}_1)^r$; $c_3 \leftarrow (X_2^t \widetilde{X}_2)^r$; $c_4 = mX_1^r$ return (c_1, c_2, c_3, c_4)
$\mathsf{decrypt}([c_1, c2, c3, c4], t)$	: $b \leftarrow [(c_2 = c_1^{tx_1 + \widetilde{x}_1}) \wedge (c_3 = c_1^{tx_2 + \widetilde{x}_2})]$ if $b = 0$ return $(\perp)$ else return $(c_4 c_1^{-x_1})$

Fig. 8.4 Cash et al.'s tag-based encryption (Cash et al. 2009)

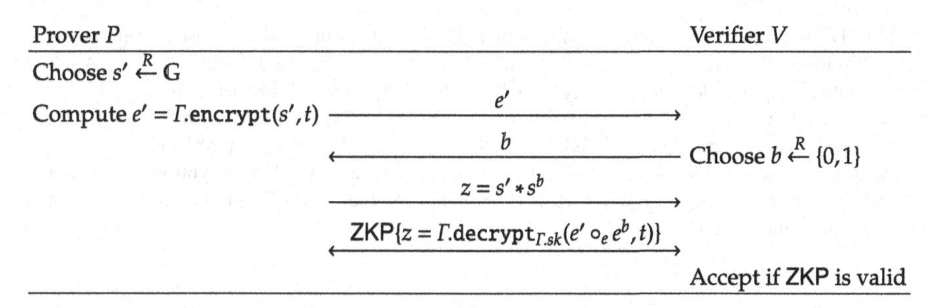

Fig. 8.5 Proof of knowledge of a decryption: $\mathsf{PoK}\{s\colon s = \Gamma.\mathtt{decrypt}_{\Gamma.sk}(e,t)\}$ **Common input:** $(e, \Gamma.pk)$ and **Private input:** s and $\Gamma.sk$ or randomness encrypting s in e

the result, described in Fig. 8.5, is an efficient protocol for proving knowledge of the decryption of a ciphertext generated using a scheme from the above-mentioned class. □

We are now ready to assess to performances of the extended paradigms in the multi-user setting.

For StE, it is natural to instantiate the signature scheme from the already mentioned class $\mathbb{S}$ (Definition 4.2), and the tag-based encryption from class $\mathbb{T}$. The confirmation/denial protocols will be a composition of the protocol in Fig. 8.5 and that depicted in Fig. 4.2.

Similarly, for CtEtS and EtS, we use the protocol depicted in Fig. 8.5 instead of that in Fig. 4.3 in the confirmation/denial protocols.

Finally, for conversion in EtS, we use the same techniques described in Sect. 6.2.3 for obtaining a non-interactive variant of the confirmation protocol.

Bottom line is that the price to pay for preserving security, namely insider unforgeability and outsider privacy, in the multi-user setting amounts to substituting IND-CPA secure encryption by IND-st-wCCA secure tag-based encryption. Practically speaking, if we consider tag-based encryption schemes from Kiltz (2006) and Cash et al. (2009) (both are IND-st-wCCA secure), the confirmer signature (signcryption) overhead is at most two group elements, whereas the computation overhead amounts to few group operations.

References

An JH, Dodis Y, Rabin T (2002) On the security of joint signature and encryption. In: Knudsen LR (ed) Advances in cryptology - EUROCRYPT 2002. LNCS, vol 2332. Springer, Heidelberg, pp 83–107

Cash D, Kiltz E, Shoup V (2009) The twin Diffie-Hellman problem and applications. J Cryptol 22(4):470–504. Earlier version in EUROCRYPT 2008

Chiba D, Matsuda T, Schuldt JN, Matsuura K (2011) Efficient generic constructions of signcryption with insider security in the multi-user setting. In: Lopez J, Tsudik G (eds) Applied cryptography and network security. LNCS, vol 6715. Springer, Heidelberg, pp 220–237

Kiltz E (2006) Chosen-ciphertext security from tag-based encryption. In: Halevi S, Rabin T (eds) Theory of cryptography (TCC 2006), vol 3876. Springer, Heidelberg, pp 581–600

Matsuda T, Matsuura K, Schuldt J (2009) Efficient constructions of signcryption schemes and signcryption composability. In: Roy B, Sendrier N (eds) IndoCrypt, vol 5922. Springer, Berlin/Heidelberg, pp 321–342

Chapter 9
Insider Privacy

Abstract So far, privacy (invisibility/confidentiality) of the case-study primitives was considered in the outsider model. This explains the success in "amplifying" security of the "base" encryption; i.e. building from CPA secure encryption (wCCA secure encryption in the multi-user setting) CCA secure confirmer signatures or signcryption. Such an amplification cannot hold in the insider model since the adversary is given the signing key and can compute valid confirmer signatures/signcryptions on messages of his choosing and then submit them for verification/decryption. Therefore, the best and most optimistic result we can hope for is to at least preserve security and base the CCA security of our primitives on the CCA security of the underlying encryption.

In addition to the expensive cost of CCA secure encryption, another caveat consists in impeding verifiability as (partially) homomorphic encryption is no longer allowed in the design. It is therefore imperative to look for an alternative encryption that allows to efficiently prove knowledge of a decryption while enjoying CCA security. In this chapter, we investigate the methods used to upgrade security in public-key encryption, and adapt them to design insider-secure confirmer signatures and signcryptions without compromising verifiability.

9.1 The CHK Transform

There are two main generic constructions of CCA secure encryption from weaker encryption. The first, due to Sahai (Sahai 1999), builds IND-CCA secure encryption from IND-CPA secure one using a non-interactive (NIZK) proof with simulation soundness. This approach suffers the high cost incurred by the use of the simulation-sound NIZK, e.g. proof size and verification cost, in addition to the necessity of a trusted authority to generate the common reference string (crs). The second conversion, referred to as the CHK transform, transforms wCCA secure tag-based encryption into CCA secure public-key encryption using a strongly unforgeable one-time signature. This is the approach we will adhere to, in this chapter, to design insider secure confirmer signatures/signcryptions.

L. El Aimani, *Verifiable Composition of Signature and Encryption*,
https://doi.org/10.1007/978-3-319-68112-2_9

9.1.1 The CHK Transform for PKE

Canetti et al. (2004) provide a method that transforms any selective-identity chosen-plaintext secure identity-based scheme into one with full-fledged chosen-ciphertext security. The transformation, referred to as the *CHK transform*, consists in signing the ciphertext, result of encryption with the weakly secure identity-based encryption scheme, using a one-time signature scheme, wherein the "identity" is given by the verification key. Later in Boneh and Katz (2005), Boneh and Katz improve the CHK transform using a MAC instead of a one-time signature; see also Boneh et al. (2007). Concurrently, MacKenzie et al. (2004) present a method for converting a weakly chosen-ciphertext secure tag-based encryption scheme to a fully secure public-key encryption scheme. Finally, Kiltz (2006) combines the ideas of Canetti et al. (2004), MacKenzie et al. (2004), and Boneh and Katz (2005) in order to derive chosen-ciphertext secure public-key encryption schemes from selective-tag weakly chosen-ciphertext secure tag-based encryption schemes using one-time signatures.

The *CHK transform*, depicted in Fig. 9.1 builds an IND-CCA secure encryption $\Gamma^\star$ from an IND-st-wCCA secure tag-based encryption Γ and a strongly unforgeable one-time signature ots.

Theorem 9.1 *The encryption scheme $\Gamma^\star$ obtained from the conversion in Fig. 9.1 is* IND-CCA *secure if Γ is* IND-st-wCCA *secure and* ots *is a strongly unforgeable one-time signature.*

Proof Let $\mathcal{A}$ be an (t, ϵ, q_d)-IND-CCA adversary against $\Gamma^\star$. We build from $\mathcal{A}$ an IND-st-wCCA adversary $\mathcal{R}$ against the underlying Γ as follows.

[keygen] $\mathcal{R}$ gets the parameters of Γ from her challenger. She further chooses a suitable (t, ϵ') strongly unforgeable one-time signature scheme ots and generates a key pair (ots.pk, ots.sk). Finally, $\mathcal{R}$ presents ots.pk to her challenger as a challenge tag. After she gets the public key pk of Γ, she forwards it, along with the parameters of Γ and ots to $\mathcal{A}$. Clearly, this simulation is perfectly indistinguishable from that in a real IND-CCA game.

[decrypt] $\mathcal{R}$ simulates the decryption oracle on queries $c_i^\star$ as follows. First parse $c_i^\star$ as (c_i, vk_i, σ_i). If ots.verify$_{vk_i}(\sigma_i, c_i) = 0$, then respond with $\perp$. Otherwise, two cases manifest: either $vk_i \neq$ ots.pk or not. In the first case, $\mathcal{R}$ requests the

$\Gamma^\star$.encrypt$_{pk}(m)$

1. (ots.pk,ots.sk) $\leftarrow$ ots.keygen(1^κ)
2. $c \leftarrow \Gamma$.encrypt$_{pk}(m$,ots.$pk)$
3. $\sigma \leftarrow$ ots.sign$_{\text{ots}.sk}(c)$
4. return (c,ots.pk,σ)

$\Gamma^\star$.decrypt$_{sk}(c^\star)$

1. Parse $c^\star$ as (c,ots.pk,σ)
2. if ots.verify$_{\text{ots}.pk}(\sigma,c) = 0$ return ($\perp$)
3. else return (Γ.decrypt$_{sk}(c$,ots.pk))

Fig. 9.1 The CHK transform

decryption of (c_i, vk_i) from her challenger whose answer will be forwarded to $\mathcal{A}$. In the second case $\mathcal{R}$ aborts the simulation.

Let F_1 be the event corresponding to $\mathcal{A}$ submitting a valid ciphertext $c_i^\star = (c_i, vk_i, \sigma_i)$, where $vk_i = \mathsf{ots}.pk$. We have $\Pr[F_1] \leq \epsilon'$, since $\mathcal{R}$ can output (c_i, σ_i) as a forgery on ots.

Therefore the probability that the provided simulation is indistinguishable from that in a standard indistinguishability game is at least $(1 - \epsilon')^{q_d}$.

[Challenge] After $\mathcal{A}$ outputs m_0 and m_1 as challenge messages to $\mathcal{R}$, the latter forwards them to her challenger. She gets in response a ciphertext c and is asked to determine the underlying message (m_0 or m_1). To do so, $\mathcal{R}$ produces a one-time signature σ on c with $\mathsf{ots}.sk$, then, she hands the triple $c^\star = (c, \mathsf{ots}.pk, \sigma)$ to $\mathcal{A}$. By construction, $c^\star$ is a valid IND-CCA challenge for $\Gamma^\star$.

[Post challenge] Consider the decryption query $c_i^\star = (c_i, vk_i, \sigma_i)$. Naturally $c_i^\star \neq c^\star$. We distinguish two types of queries. If $vk_i \neq \mathsf{ots}.pk$, then $\mathcal{R}$ proceeds as usual using her own decryption oracle. Otherwise, $\mathcal{R}$ responds with $\perp$ if the output of $\mathsf{ots}.\mathtt{verify}_{vk_i}(\sigma_i, c_i)$ is 0, and aborts otherwise.

Let F_2 be the event corresponding to $\mathcal{A}$ submitting a valid ciphertext (c_i, vk_i, σ_i) where $vk_i = \mathsf{ots}.pk$. By definition of one-time signatures, $c_i = c$, since otherwise, $\mathcal{A}$ would be able to forge a valid one-time signature on a new message, namely c_i, w.r.t. $\mathsf{ots}.pk$. Now, if $c_i = c$, then $\sigma \neq \sigma_i$ (as $c_i^\star \neq c^\star$). It follows that (c_i, σ_i) is a strong forgery on ots.

Similarly, this simulation is indistinguishable from that in a real IND-CCA game by at least $(1 - \epsilon')^{q_d}$.

Finally, when $\mathcal{A}$ answers b, $\mathcal{R}$ will return the same answer to her challenger. $\mathcal{R}$ will succeed in solving her challenge with advantage at least $\epsilon(1 - \epsilon')^{q_d}$.

□

9.1.2 A CHK-Like Transform for TBE

In this paragraph, we propose a transformation to upgrade security in TBE, namely to build fully secure (IND-CCA) TBE from IND-st-wCCA secure TBE.

In Kiltz (2006), Kiltz suggests to achieve this task through the intermediary of a CCA secure public-key encryption (PKE): first derive a CCA secure PKE from a st-wCCA secure TBE using the conversion depicted in Fig. 9.1, then identify the pair "(message,tag)" in the TBE by the new message "message||tag" in the PKE.

The transform described in Fig. 9.2, first introduced in El Aimani and Joye (2013) in the context of group encryption, has the merit of preserving the algebraic structure of the message to be encrypted. This impacts positively the verifiability of the encryption, i.e. proving knowledge of the message underlying the CCA encryption is as efficient as proving knowledge of the message underlying its st-wCCA encryption.

$\Gamma^\star$.encrypt$_{pk}(m,L)$	$\Gamma^\star$.decrypt$_{sk}(c^\star,L)$
1. $(\text{ots}.pk, \text{ots}.sk) \leftarrow \text{ots.keygen}(1^\kappa)$	1. Parse $c^\star$ as $(c, \text{ots}.pk, \sigma)$
2. $c \leftarrow \Gamma.\text{encrypt}_{pk}(m, \text{ots}.pk)$	2. If $\text{ots.verify}_{\text{ots}.pk}(\sigma, c\|L) = 0$ return $(\perp)$
3. $\sigma \leftarrow \text{ots.sign}_{\text{ots}.sk}(c\|L)$	3. else return $(\Gamma.\text{decrypt}_{sk}(c, \text{ots}.pk))$
4. return $(c, \text{ots}.pk, \sigma)$	

Fig. 9.2 The CHK-like transform for TBE

Theorem 9.2 *The tag-based encryption scheme $\Gamma^\star$ obtained from the conversion in Fig. 9.2 is* IND-CCA *secure if Γ is* IND-st-wCCA *secure and* ots *is a strongly unforgeable one-time signature.*

Proof (Sketch) The proof is similar to that of Theorem 9.1 so we avoid the redundancies and focus on the differences.

Simulation queries before the challenge phase and construction of the challenge are handled as in the proof of Theorem 9.1.

Let $(c^\star, L) = [(c, \text{ots}.pk, \sigma), L]$ be the challenge handed to the adversary. Note that L is the challenge label presented by the adversary before the reduction, whereas ots.pk forms the challenge label presented by the reduction to her own challenger.

Let further $q_i = (c_i^\star, L_i) = [(c_i, vk_i, \sigma_i), L_i]$ be a decryption query issued to the reduction upon receipt of the challenge. Naturally $(c_i^\star, L_i) \neq (c^\star, L)$. We distinguish two types of queries. If $vk_i \neq \text{ots}.pk$, then the reduction proceeds as usual using her decryption oracle. Otherwise she responds with $\perp$ if σ_i is not a valid signature on $c_i\|L_i$ (w.r.t. vk_i), and aborts otherwise.

Let F denote the event corresponding to the adversary submitting a valid ciphertext $c_i^\star$, w.r.t. label L_i, where $vk_i = \text{ots}.pk$. We distinguish two cases:

1. If $L_i = L$, and thus $c_i^\star \neq c^\star$ ($q_i \neq (c^\star, L)$), then $(c_i, \sigma_i) \neq (c, \sigma)$ and thus $(c_i\|L_i, \sigma_i) \neq (c\|L, \sigma)$, which means that $(c_i\|L_i, \sigma_i)$ is a strong forgery on ots.
2. If $L_i \neq L$, then $c_i\|L_i \neq c\|L$. It follows that $(c_i\|L_i, \sigma_i) \neq (c\|L, \sigma)$, and thus $(c_i\|L_i, \sigma_i)$ is a strong forgery on ots.

We conclude that the reduction is able to correctly simulate the decryption queries provided the one-time signature is strongly unforgeable. Finally, when the adversary solves the IND-CCA challenge, the reduction will forward this very answer to her IND-st-wCCA challenger.

9.2 New Paradigms with Insider Privacy

We are now ready to describe the extended StE, CtEaS, and EtS paradigms that proffer an insider privacy while enjoying efficient verifiability. The key ideas behind the new constructions are:

- For each run of the signing (signcryption) procedure, produce through the one-time signature key generation algorithm, a pair of signing and verification keys (sk, vk).
- Use tag-based encryption whenever encrypting anything, where the tag is set to the freshly generated verification key vk of the one-time signature.
- Whenever signing, include in addition to the message to be signed, the verification key vk of the one-time signature.
- Generate, using sk, a one-time signature at the outer layer.

While the use of tag-based encryption combined with one-time signature has been justified earlier, including the one-time signature verification key in the message to be signed deserves a bit of explanation.

Recall that the plain StE and CtEaS suffer the strong forgeability, even if the underlying signature is strongly unforgeable; an adversary can easily forge on a queried message by re-encrypting the digital signature (and the message) in case of StE, and the commitment string (and the message) in case of CtEaS. This shortcoming has been addressed by binding the signature to the encryption: for StE, the digital signature is produced on the message and the encapsulation of the key that will later be used for encrypting the signature (see Chap. 4 or Chiba et al. 2011; El Aimani 2009), whereas for CtEaS, the signature is produced on the commitment and the encryption of the commitment string (see Chap. 6 or El Aimani 2010). With these fixes, the strong unforgeability is preserved in constructions from StE and CtEaS.

Recently, Nandi and Pandit (2016) remarked that the use of one-time signatures brings as a by-product amplification of the signature security. Let's check closely the StE case for confirmer signatures, the other cases being similar.

Let $\mu = (c, \mathsf{ots}.vk, \mathsf{ots}.\mathtt{sign}_{\mathsf{ots}.sk}(c))$ be a confirmer signature on some message m, where c is encryption w.r.t. tag $\mathsf{ots}.vk$ of some digital signature s on $m \| \mathsf{ots}.vk$. Since we assume the existential unforgeability of the base signature, the best that an unforgeability adversary can come up with is another digital signature s' on $m \| \mathsf{ots}.vk$, that he encrypts in c'. However, to be able to forge a new confirmer signature on m, the adversary must compute a one-time signature on the newly computed ciphertext c' using the private key corresponding to $\mathsf{ots}.vk$; this is computationally impossible thanks to the strong unforgeability of the one-time signature. In other words, the strong unforgeability of the one-time signature compensates for the weak unforgeability of the base signature, if the latter is cautiously applied on the message to be signed and the verification key of the one-time signature.

9.2.1 Constructions for Confirmer Signatures

In addition to the classical building blocks, namely, a signature scheme Σ, an encryption scheme Γ that supports labels, and a commitment scheme Ω, we further need a one-time signature ots.

To simplify the exposition of the new paradigms, we omit, in the verification/conversion procedures, verification of the validity of the one-time signature performed at the outer layer. In fact, if the latter is invalid, then verification/conversion would obviously return $\perp$. The same thing applies to CtEaS or EtS in case the digital signature is invalid w.r.t. the given commitment or ciphertext respectively.

Finally, recall that in the StE paradigm, signatures generated using the digital signature scheme Σ can be efficiently transformed in a reversible way into a pair (s, r) where s is the "significant" part of the signature and r reveals no information about the signed message or $(\Sigma.sk, \Sigma.pk)$.

Remark 9.1 Note that the encryption scheme Γ need not be derived from the hybrid encryption (KEM/DEM) paradigm. In fact, strong unforgeability is ensured thanks to the use of one-time signatures, as mentioned previously.

Theorem 9.3 (StE with Insider Privacy) *The StE paradigm, described in Fig. 9.3 is* SEUF-CMA *secure if the underlying signature scheme is* EUF-CMA *secure and the used one-time signature is strongly unforgeable. Moreover, it is* INV-CMA *secure, in the* insider model, *if it uses an* IND-st-wCCA *secure tag-based encryption and a strongly unforgeable one-time signature.*

$\mathtt{setup}(1^\kappa)$	: $\Sigma.\mathtt{setup}(1^\kappa)$; $\Gamma.\mathtt{setup}(1^\kappa)$; $\mathsf{ots.setup}(1^\kappa)$
$\mathtt{keygen_S}(1^\kappa)$	: $\Sigma.\mathtt{keygen}(1^\kappa)$
$\mathtt{keygen_C}(1^\kappa)$	: $\Gamma.\mathtt{keygen}(1^\kappa)$
$\mathtt{sign}(m)$	: $(sk,vk) \leftarrow \mathsf{ots.keygen}(1^\kappa)$; $(s,r) \leftarrow \Sigma.\mathtt{sign}_{\Sigma.sk}(m\|vk)$
	$c \leftarrow \Gamma.\mathtt{encrypt}_{\{\Gamma.pk,coins\}}(s,vk)$; $\tau = \mathsf{ots.sign}_{sk}(c\|r)$
	return(c,r,vk,τ)
$\mathtt{sconfirm}([c,r,vk,\tau],m)$	: $\mathsf{ZKP}\{(s,coins): c = \Gamma.\mathtt{encrypt}_{\{\Gamma.pk,coins\}}(s,vk) \wedge$
	$\Sigma.\mathtt{verify}_{\Sigma.pk}([s,r],m\|vk) = 1\}$
$\mathtt{confirm}([c,r,vk,\tau],m)$	: $\mathsf{ZKP}\{(s,\Gamma.sk): s = \Gamma.\mathtt{decrypt}_{\{\Gamma.sk\}}(c,vk) \wedge$
	$\Sigma.\mathtt{verify}_{\Sigma.pk}([s,r],m\|vk) = 1\}$
$\mathtt{deny}([c,r,vk,\tau],m)$	: $\mathsf{ZKP}\{(s,\Gamma.sk): s = \Gamma.\mathtt{decrypt}_{\{\Gamma.sk\}}(c,vk) \wedge$
	$\Sigma.\mathtt{verify}_{\Sigma.pk}([s,r],m\|vk) = 0\}$
$\mathtt{convert}([c,r,vk,\tau],m)$	: $s \leftarrow \Gamma.\mathtt{decrypt}_{\Gamma.sk}(c,vk)$; $b \leftarrow \Sigma.\mathtt{verify}_{\Sigma.pk}([s,r],m\|vk)$
	if $b = 0$ return $(\perp)$ else return (s,r,vk)

Fig. 9.3 StE with insider privacy

Proof (Sketch) Unforgeability is straightforward.

For invisibility, we focus only on the differences with the outsider model and avoid the redundancies.

Simulation of the signature, verification, and conversion queries is straightforward using the decryption oracle wCCA and the strong unforgeability of the one-time signature (see proof of Theorem 9.1).

Once the adversary outputs the challenge messages, say, m_0 and m_1, the reduction produces digital signatures (s_0, r_0) and (s_1, r_1) on $m_0\|vk$ and $m_1\|vk$ respectively (vk is the challenge tag the reduction presented to her challenger), and presents (s_0, s_1) to her own challenger. She gets in response an encryption c_b of s_b for some $b \in \{0, 1\}$. To build her challenge confirmer signature, she chooses a bit $b' \overset{R}{\leftarrow} \{0, 1\}$ and computes a one-time signature τ on $c_b\|r_{b'}$. Finally she hands $\mu = (c_b, r_{b'}, vk, \tau)$.

Two cases, either $b = b'$ in which case μ is a valid confirmer signature on m_b, or $b \neq b'$ and thus μ invalid w.r.t. either message (we assume that $r_0 \neq r_1$, otherwise μ_b is trivially a valid signature on m_b regardless of b). The only way to distill information about the message underlying μ is c_b as $r_{b'}$ is by assumption statistically independent of both messages. Therefore, when the adversary outputs his guess, say b'', the reduction will forward it to her own challenger if $b'' = b'$, otherwise she forwards $1 - b''$. □

Remark 9.2 Note that the new CtEaS, outlined in Fig. 9.4, regained the parallelism between signature and encryption enjoyed by the original paradigm. In fact, strong unforgeability is now guaranteed thanks to the one-time signature, as mentioned previously.

setup(1^κ)	$:\{\Sigma, \Gamma, \Omega, \mathsf{ots}\}.\mathtt{setup}(1^\kappa)$
keygen$_S(1^\kappa)$	$: \Sigma.\mathtt{keygen}(1^\kappa)$
keygen$_C(1^\kappa)$	$: \Gamma.\mathtt{keygen}(1^\kappa)$
sign(m)	$: (sk, vk) \leftarrow \mathsf{ots.keygen}(1^\kappa)\ ;\ c \leftarrow \Omega.\mathtt{commit}(m,r)$
	$e \leftarrow \Gamma.\mathtt{encrypt}_{\{\Gamma.pk,coins\}}(r,vk)\ ;\ \sigma \leftarrow \Sigma.\mathtt{sign}_{\Sigma.sk}(c\|vk)$
	$\tau \leftarrow \mathsf{ots.sign}_{sk}(c\|e\|\sigma)\ ;\ \mathsf{return}\ (c,e,\sigma,vk,\tau)$
sconfirm($[c,e,\sigma,vk,\tau]$,m)	$: \mathsf{ZKP}\big\{(r,coins) : c = \Omega.\mathtt{commit}(m,r)\ \wedge$
	$e = \Gamma.\mathtt{encrypt}_{\{\Gamma.pk,coins\}}(r,vk)\big\}$
confirm($[c,e,\sigma,vk,\tau]$,m)	$: \mathsf{ZKP}\big\{(r,\Gamma.sk) : c = \Omega.\mathtt{commit}(m,r)\ \wedge$
	$r = \Gamma.\mathtt{decrypt}_{\{\Gamma.sk\}}(e,vk)\big\}$
deny($[c,e,\sigma,vk,\tau]$,m)	$: \mathsf{ZKP}\big\{(r,\Gamma.sk) : c \neq \Omega.\mathtt{commit}(m,r)\ \wedge$
	$r = \Gamma.\mathtt{decrypt}_{\{\Gamma.sk\}}(e,vk)\big\}$
convert($[c,e,\sigma,vk,\tau]$,m)	$: r \leftarrow \Gamma.\mathtt{decrypt}_{\Gamma.sk}(e,vk)\ ;\ b \leftarrow (c = \Omega.\mathtt{commit}(m,r))$
	if $b = 0$ return $(\perp)$ else return (r,c,e,vk,σ)

Fig. 9.4 CtEaS with insider privacy

setup(1^κ)	: $\{\Sigma, \Gamma, \text{ots}\}$.setup($1^\kappa$) ; crs $\leftarrow$ TA.setup(1^κ)
keygen$_S$(1^κ)	: Σ.keygen(1^κ)
keygen$_C$(1^κ)	: Γ.keygen(1^κ)
sign(m)	: $(sk, vk) \leftarrow$ ots.keygen(1^κ) ; $c \leftarrow \Gamma.\text{encrypt}_{\{\Gamma.pk, coins_c\}}(m, vk)$
	$\sigma \leftarrow \Sigma.\text{sign}_{\Sigma.sk}(c\|vk)$; $\tau \leftarrow \text{ots.sign}_{sk}(c\|\sigma)$
	return $[c, \sigma, vk, \tau]$
sconfirm($[c, \sigma, vk, \tau]$,m)	: ZKP$\left\{coins_c \colon c = \Gamma.\text{encrypt}_{\{\Gamma.pk, coins_c\}}(m, vk)\right\}$
confirm($[c, \sigma, vk, \tau]$,m)	: ZKP$\{\Gamma.sk \colon m = \Gamma.\text{decrypt}_{\Gamma.sk}(c, vk)\}$
deny($[c, \sigma, vk, \tau]$,m)	: ZKP$\{\Gamma.sk \colon m \neq \Gamma.\text{decrypt}_{\Gamma.sk}(c, vk)\}$
convert($[c, \sigma, vk, \tau]$,m)	: $\pi \leftarrow$ NIZK$\{m = \Gamma.\text{decrypt}_{\Gamma.sk}(c, vk)\}$; return (π, c, vk, σ)

Fig. 9.5 EtS with insider privacy

Theorem 9.4 (CtEaS with Insider Privacy) *The CtEaS paradigm, described in Fig. 9.4 is* SEUF-CMA *secure if the underlying signature scheme is* EUF-CMA *secure, the used commitment is computationally binding, and the used one-time signature is strongly unforgeable. Moreover, it is* INV-CMA *secure,* in the insider model, *if it uses a computationally hiding commitment, an* IND-st-wCCA *secure tag-based encryption, and a strongly unforgeable one-time signature.* □

Theorem 9.5 (EtS with Insider Privacy) *The EtS paradigm, described in Fig. 9.5 is* SEUF-CMA- *secure if the underlying signature scheme is* EUF-CMA *secure and the used one-time signature is strongly unforgeable. Moreover, it is* INV-CMA *secure,* in the insider model, *if it uses an* IND-st-wCCA *secure tag-based encryption and a strongly unforgeable one-time signature.* □

9.2.2 *Constructions for Verifiable Signcryption*

Recall that only EtS provides efficient verifiable signcryption. The paradigm EtS in Fig. 9.5 gives also secure verifiable signcryptions in the insider model (SEUF-CMA and IND-CCA) if the conditions on the building blocks are met as outlined in Theorem 9.5.

Since EtS fails to lavish anonymity of the sender as it exposes the signing public key, we have proposed an alternative—the EtStE paradigm—that overcomes this shortcoming. In Fig 9.6, we outline an extension of this paradigm that is secure in the insider model (strong unforgeability SEUF-CMA and full indistinguishability IND-CCA) if it uses an EUF-CMA secure signature Σ, IND-st-wCCA secure tag-based encryption (both encryption schemes Γ_1 and Γ_2) and a strongly unforgeable one-time signature ots.

Similarly to the outsider secure variant, EtStE uses a signature scheme assumed to issue signatures that can be written in a reversible way as a pair (s, r), where s

setup(1^κ)	: $\{\Sigma,\Gamma_1,\Gamma_2,\mathsf{ots}\}$.setup($1^\kappa$); crs $\leftarrow$ TA.setup(1^κ)
keygen$_S$(1^κ)	: Σ.keygen(1^κ)
keygen$_R$(1^κ)	: $\{\Gamma_1,\Gamma_2\}$.keygen(1^κ)
signcrypt(m)	: $(sk,vk) \leftarrow \mathsf{ots.keygen}(1^\kappa)$; $c_1 \leftarrow \Gamma_1.\mathtt{encrypt}_{\{\Gamma_1.pk,coins_1\}}(m,vk)$
	$(s,r) \leftarrow \Sigma.\mathtt{sign}_{\Sigma.sk}(c_1\|vk)$; $c_2 \leftarrow \Gamma_2.\mathtt{encrypt}_{\{\Gamma_2.pk,coins_2\}}(s,vk)$
	$\tau \leftarrow \mathsf{ots.sign}_{sk}(c_1\|c_2\|r)$; return $[c_1,c_2,r,vk,\tau]$
proveValidity(c_1,c_2,r,vk,τ)	: $\mathsf{ZKP}\{(m,coins_1,s,coins_2)\colon c_1 = \Gamma_1.\mathtt{encrypt}_{\{\Gamma_1.pk,coins_1\}}(m,vk) \wedge$
	$c_2 = \Gamma_2.\mathtt{encrypt}_{\{\Gamma_2.pk,coins_2\}}(s,vk) \wedge$
	$\Sigma.\mathtt{verify}_{\Sigma.pk}([s,r],c_1\|vk) = 1 \}$
unsigncrypt(c_1,c_2,r,vk,τ)	: $s \leftarrow \Gamma_2.\mathtt{decrypt}_{\Gamma_2.sk}(c_2,vk)$; $b \leftarrow \Sigma.\mathtt{verify}_{\Sigma.pk}([s,r],c_1\|vk)$
	if $b=0$ return ($\perp$) else return ($\Gamma_1.\mathtt{decrypt}_{\Gamma_1.sk}(c_1,vk)$)
confirm([c_1,c_2,r,vk,τ],m)	: $\mathsf{ZKP}\{(s,\Gamma_1.sk,\Gamma_2.sk)\colon m = \Gamma_1.\mathtt{decrypt}_{\Gamma_1.sk}(c_1,vk) \wedge$
	$s = \Gamma_2.\mathtt{decrypt}_{\{\Gamma_2.sk\}}(c_2,vk) \wedge$
	$\Sigma.\mathtt{verify}_{\Sigma.pk}([s,r],c_1\|vk) = 1 \}$
deny([c_1,c_2,r,vk,τ],m)	: $\mathsf{ZKP}\{\Gamma_1.sk\colon m \neq \Gamma_1.\mathtt{decrypt}_{\Gamma_1.sk}(c_1,vk)\} \vee$
	$\mathsf{ZKP}\{(s,\Gamma_2.sk)\colon s = \Gamma_2.\mathtt{decrypt}_{\{\Gamma_2.sk\}}(c_2,vk) \wedge$
	$\Sigma.\mathtt{verify}_{\Sigma.pk}([s,r],c_1\|vk) \neq 1 \}$
sigExtract([c_1,c_2,r,vk,τ],m)	: $s \leftarrow \Gamma_2.\mathtt{decrypt}_{\Gamma_2.pk}(c_2,vk)$; $b \leftarrow \Sigma.\mathtt{verify}_{\Sigma.pk}([s,r],c_1\|vk)$
	if $b=0$ return ($\perp$) else
	$\{\pi \leftarrow \mathsf{NIZK}\{m = \Gamma.\mathtt{decrypt}_{\Gamma_1.sk}(c_1,vk)\}$; return $[\pi,c_1,s,r,vk]\}$
sigVerify([π,c_1,s,r,vk],m)	: NIZK.verify(crs,π) ; $\Sigma.\mathtt{verify}_{\Sigma.pk}([s,r],c_1\|vk)$

Fig. 9.6 EtStE with insider privacy

represents the "significant" part of the signature, and r reveals no information about the signed message nor about the public signing key.

Note also that Γ_2 need not be derived from the KEM/DEM paradigm. In fact, the strong unforgeability of the one-time signature ensures the strong unforgeability of the whole construction if the used digital signature is unforgeable.

Theorem 9.6 (EtStE with Insider Privacy) *The EtStE paradigm, described in Fig. 9.6 is* SEUF-CMA- *secure if the underlying signature scheme is* EUF-CMA *secure and the used one-time signature is strongly unforgeable. Moreover, it is* INV-CMA *secure,* in the insider model, *if it uses* IND-st-wCCA *secure tag-based encryption (both Γ_1 and Γ_2) and a strongly unforgeable one-time signature.* □

9.2.3 Multi-User Security

Multi-user security in the insider model can be achieved using the same tricks used in the outsider model, namely add the public key of the confirmer (receiver) to the

$\mathsf{setup}(1^\kappa)$	$: \Sigma.\mathsf{setup}(1^\kappa) \,;\, \Gamma.\mathsf{setup}(1^\kappa) \,;\, \mathsf{ots.setup}(1^\kappa)$
$\mathsf{keygen}_S(1^\kappa)$	$: \Sigma.\mathsf{keygen}(1^\kappa)$
$\mathsf{keygen}_C(1^\kappa)$	$: \Gamma.\mathsf{keygen}(1^\kappa)$
$\mathsf{sign}(m, \Sigma.pk, \Gamma.pk)$	$: (sk, vk) \leftarrow \mathsf{ots.keygen}(1^\kappa)$ $(s,r) \leftarrow \Sigma.\mathsf{sign}_{\Sigma.sk}(m \Vert \Gamma.pk \Vert vk)$ $c \leftarrow \Gamma.\mathsf{encrypt}_{\{\Gamma.pk, coins\}}(s, vk)$ $\tau = \mathsf{ots.sign}_{sk}(c \Vert r \Vert \Sigma.pk) \,;\, \text{return } (c, r, vk, \tau)$
$\mathsf{sconfirm}([c,r,vk,\tau], m, \Sigma.pk, \Gamma.pk)$	$: \mathsf{ZKP}\{(s, coins) : c = \Gamma.\mathsf{encrypt}_{\{\Gamma.pk, coins\}}(s, vk) \;\wedge$ $\Sigma.\mathsf{verify}_{\Sigma.pk}([s,r], m \Vert \Gamma.pk \Vert vk) = 1\}$
$\mathsf{confirm}([c,r,vk,\tau], m, \Sigma.pk, \Gamma.pk)$	$: \mathsf{ZKP}\{(s, \Gamma.sk) : s = \Gamma.\mathsf{decrypt}_{\{\Gamma.sk\}}(c, vk) \;\wedge$ $\Sigma.\mathsf{verify}_{\Sigma.pk}([s,r], m \Vert \Gamma.pk \Vert vk) = 1\}$
$\mathsf{deny}([c,r,vk,\tau], m, \Sigma.pk, \Gamma.pk)$	$: \mathsf{ZKP}\{(s, \Gamma.sk) : s = \Gamma.\mathsf{decrypt}_{\{\Gamma.sk\}}(c, vk) \;\wedge$ $\Sigma.\mathsf{verify}_{\Sigma.pk}([s,r], m \Vert \Gamma.pk \Vert vk) = 0\}$
$\mathsf{convert}([c,r,vk,\tau], m, \Sigma.pk, \Gamma.pk)$	$: s \leftarrow \Gamma.\mathsf{decrypt}_{\Gamma.sk}(c, vk)$ if $\Sigma.\mathsf{verify}_{\Sigma.pk}([s,r], m \Vert \Gamma.pk \Vert vk) = 0$ return $(\perp)$ else return (s, r, vk)

Fig. 9.7 Multi-user StE with insider privacy

message to be signed, and whenever encrypting, use a tag-based encryption scheme, where the tag is set to the public key of the signer (sender).

Since we opted for tag-based encryption combined with secure one-time signature to achieve insider privacy without compromising verifiability, we maintain the same design with the extra step of implementing the above tips to get multi-user security. As a consequence, the public key of the signer (sender) is added, as per the transform in Fig. 9.2, to the components to be one-time signed by ots. Finally, recall that the freshly generated verification key vk of ots is added to the signed message in order to supplement the constructions with strong unforgeability.

Theorem 9.7 *The paradigms, described in Figs. 9.7, 9.8, and 9.9, preserve the security proven in Theorems 9.3, 9.4, and 9.5 respectively, in the multi-user setting.* □

9.2.4 Performance

The compositions proposed in this section, either in the two-user or multi-user setting, entail the use of IND-st-wCCA secure encryption and one-time signatures, in addition to the classical signature and commitment schemes. Besides, verification (and conversion in case of EtS) comprises always the already mentioned proofs, namely:

$\mathtt{setup}(1^\kappa)$	: $\{\Sigma, \Gamma, \Omega, \mathsf{ots}\}.\mathtt{setup}(1^\kappa)$
$\mathtt{keygen_S}(1^\kappa)$	: $\Sigma.\mathtt{keygen}(1^\kappa)$
$\mathtt{keygen_C}(1^\kappa)$	: $\Gamma.\mathtt{keygen}(1^\kappa)$
$\mathtt{sign}(m, \Sigma.pk, \Gamma.pk)$	: $(sk, vk) \leftarrow \mathsf{ots.keygen}(1^\kappa)$; $c \leftarrow \Omega.\mathtt{commit}(m, r)$ $e \leftarrow \Gamma.\mathtt{encrypt}_{\{\Gamma.pk, coins\}}(r, vk)$ $\sigma \leftarrow \Sigma.\mathtt{sign}_{\Sigma.sk}(c \Vert \Gamma.pk \Vert vk)$ $\tau \leftarrow \mathsf{ots.sign}_{sk}(c \Vert e \Vert \sigma \Vert \Sigma.pk)$; return (c, e, σ, vk, τ)
$\mathtt{sconfirm}([c, e, \sigma, vk, \tau], m, \Sigma.pk, \Gamma.pk)$	: $\mathsf{ZKP}\big\{(r, coins) \colon c = \Omega.\mathtt{commit}(m, r) \wedge e = \Gamma.\mathtt{encrypt}_{\{\Gamma.pk, coins\}}(r, vk)\big\}$
$\mathtt{confirm}([c, e, \sigma, vk, \tau], m, \Sigma.pk, \Gamma.pk)$	: $\mathsf{ZKP}\big\{(r, \Gamma.sk) \colon c = \Omega.\mathtt{commit}(m, r) \wedge r = \Gamma.\mathtt{decrypt}_{\{\Gamma.sk\}}(e, vk)\big\}$
$\mathtt{deny}([c, e, \sigma, vk, \tau], m, \Sigma.pk, \Gamma.pk)$	: $\mathsf{ZKP}\big\{(r, \Gamma.sk) \colon c \neq \Omega.\mathtt{commit}(m, r) \wedge r = \Gamma.\mathtt{decrypt}_{\{\Gamma.sk\}}(e, vk)\big\}$
$\mathtt{convert}([c, e, \sigma, vk, \tau], m, \Sigma.pk, \Gamma.pk)$	: $r \leftarrow \Gamma.\mathtt{decrypt}_{\Gamma.sk}(e, vk)$; $b \leftarrow (c = \Omega.\mathtt{commit}(m, r))$ if $b = 0$ return $(\perp)$ else return $\{r, c, \sigma, vk\}$

Fig. 9.8 Multi-user CtEaS with insider privacy

$\mathtt{setup}(1^\kappa)$	: $\{\Sigma, \Gamma, \mathsf{ots}\}.\mathtt{setup}(1^\kappa)$; $\mathsf{crs} \leftarrow \mathsf{TA}.\mathtt{setup}(1^\kappa)$
$\mathtt{keygen_S}(1^\kappa)$	: $\Sigma.\mathtt{keygen}(1^\kappa)$
$\mathtt{keygen_C}(1^\kappa)$	: $\Gamma.\mathtt{keygen}(1^\kappa)$
$\mathtt{sign}(m, \Sigma.pk, \Gamma.pk)$	: $(sk, vk) \leftarrow \mathsf{ots.keygen}(1^\kappa)$; $c \leftarrow \Gamma.\mathtt{encrypt}_{\Gamma.pk}(m, vk)$ $\sigma \leftarrow \Sigma.\mathtt{sign}_{\Sigma.sk}(c \Vert \Gamma.pk \Vert vk)$; $\tau \leftarrow \mathsf{ots.sign}_{sk}(c \Vert \sigma \Vert \Sigma.pk)$ return $[c, \sigma, vk, \tau]$
$\mathtt{sconfirm}([c, \sigma, vk, \tau], m, \Sigma.pk, \Gamma.pk)$	: $\mathsf{ZKP}\big\{coins_c \colon c = \Gamma.\mathtt{encrypt}_{\{\Gamma.pk, coins_c\}}(m, vk)\big\}$
$\mathtt{confirm}([c, \sigma, vk, \tau], m, \Sigma.pk, \Gamma.pk)$	: $\mathsf{ZKP}\{\Gamma.sk \colon m = \Gamma.\mathtt{decrypt}_{\Gamma.sk}(c, vk)\}$
$\mathtt{deny}([c, \sigma, vk, \tau], m, \Sigma.pk, \Gamma.pk)$	: $\mathsf{ZKP}\{\Gamma.sk \colon m \neq \Gamma.\mathtt{decrypt}_{\Gamma.sk}(c, vk)\}$
$\mathtt{convert}([c, \sigma, vk, \tau], m, \Sigma.pk, \Gamma.pk)$	: $\pi \leftarrow \mathsf{NIZK}\{m = \Gamma.\mathtt{decrypt}_{\Gamma.sk}(c, vk)\}$; return $[\pi, c, \sigma, vk]$

Fig. 9.9 Multi-user EtS with insider privacy

- Proof of knowledge of a signature on a known message. Such a proof can be easily conducted if the signature scheme belongs to class $\mathbb{S}$ (Definition 4.2).
- Proof that a commitment is correctly produced on a given message. Again, this proof can be carried out efficiently if the commitment belongs to class $\mathbb{C}$ (Definition 6.1).
- Proof of knowledge of the decryption of some ciphertext. We provided in Fig. 8.5 such a proof if the used encryption belongs to class $\mathbb{T}$ (Definition 8.3).
- Proof that a message is/isn't the decryption of some ciphertext. We explained in Sect. 6.2.2 how to conduct such proofs.

Therefore, the extra-cost, both in bandwidth and computation, induced by insider privacy in the above constructions comes from the used one-time signature. If we use the one-time signature from Groth (2006), the confirmer signature (signcryption) overhead will be five group elements, in addition to the computations incurred by the one-time signature operations.

References

Boneh D, Katz J (2005) Improved efficiency for CCA-secure cryptosystems built using identity-based encryption. In: Menezes A (ed) Topics in cryptology – CT-RSA 2005, vol 3027. Springer, Heidelberg, pp 87–103

Boneh D, Canetti R, Halevi S, Katz J (2007) Chosen-ciphertext security from identity-based encryption. SIAM J Comput 36(5):1301–1328

Canetti R, Halevi S, Katz J (2004) Chosen-ciphertext security from identity-based encryption. In: Cachin C, Camenisch J (eds) Advances in cryptology – EUROCRYPT 2004, vol 3027. Springer, Heidelberg, pp 207–222

Chiba D, Matsuda T, Schuldt JN, Matsuura K (2011) Efficient generic constructions of signcryption with insider security in the multi-user setting. In: Lopez J, Tsudik G (eds) Applied cryptography and network security. LNCS, vol 6715. Springer, Heidelberg, pp 220–237

El Aimani L (2009) On generic constructions of designated confirmer signatures. In: Roy B, Sendrier N (eds) Progress in cryptology - INDOCRYPT 2009, vol 5922. Springer, Berlin/Heidelberg, pp 343–362. Full version available at the Cryptology ePrint Archive, Report 2009/403

El Aimani L (2010) Efficient confirmer signature from the "signature of a commitment" paradigm. In: Heng SH, Kurosawa K (eds) ProvSec 2010. LNCS, vol 6402. Springer, Heidelberg, pp 87–101. Full version available at the Cryptology ePrint Archive, Report 2009/435

El Aimani L, Joye M (2013) Toward practical group encryption. In: ACNS 2013. Springer, Heidelberg, pp 237–252

Groth J (2006) Simulation-sound NIZK proofs for a practical language and constant size group signatures. In: Lai X, Chen K (eds) ASIACRYPT. LNCS, vol 4284. Springer, Heidelberg, pp 444–459

Kiltz E (2006) Chosen-ciphertext security from tag-based encryption. In: Halevi S, Rabin T (eds) Theory of cryptography (TCC 2006), vol 3876. Springer, Heidelberg, pp 581–600

MacKenzie PD, Reiter MK, Yang K (2004) Alternatives to non-malleability: definitions, constructions, and applications. In: Naor M (ed) Theory of cryptography (TCC 2004), vol 2951. Springer, Heidelberg, pp 171–190

Nandi M, Pandit T (2016) On the security of joint signature and encryption revisited. J Math Cryptol 10(3–4):181–221

Sahai A (1999) Non-malleable non-interactive zero knowledge and adaptive chosen-ciphertext security. In: Beame P (ed) Proceedings of the 40th IEEE symposium on foundations of computer science (FOCS'99). IEEE Computer Society, New York, pp 543–553

Chapter 10
Wrap-Up

Among the panoply of cryptographic applications that erupted with the electronic era, we confined ourselves to the study of applications that require integrity and authenticity of the transmitted data, in addition to confidentiality combined with the possibility of proving given properties about the hidden information. Our study was conducted in the context of two cryptographic primitives, namely confirmer signatures and signcryption. However, we expect the approach to be extended, at least partially, to further cryptographic systems that hinge upon digital signatures and verifiable encryption. In this summary, we highlight the key steps of our work that are likely to occur when studying the aforementioned systems.

Generic constructions Our study of the mentioned primitives is in line with the widespread approach adopted in cryptographic design that consists in first building "high level" or complicated systems upon basic primitives, then improving the efficiency of the resulting construction. This approach has been immensely successful as it gives easy-to-analyze and easy-to-prove systems. In fact, systems' complexity is one of the main causes of their failure, especially in hostile environments in which operate cryptographic schemes. In line with this, generic constructions benefit from the extensive research (positive or negative) that was carried out on the base primitives. They also allow to easily exchange individual system components at very little expense and effort. As a consequence, this approach produces robust systems whose efficiency compete with that of the monolithic realizations of the cryptosystems in question.

Fit-for-application security notions To anticipate attacks of variable degrees of severity, we had to consider different security notions and propose constructions accordingly. We also subjected our adhered to notions to an extensive comparison with the established security properties. In fact, some of these latter are unnecessarily stringent and do not seem to be implementable in practice; as an illustration, an insider privacy adversary who is prohibited from querying valid confirmer signatures (signcryptions) on the challenge messages.

L. El Aimani, *Verifiable Composition of Signature and Encryption*,
https://doi.org/10.1007/978-3-319-68112-2_10

Identification of design flaws A flagship phase of the conducted study was the identification of the various obstacles that hinder efficiency in the known paradigms. Actually, once these defects were spot, looking for alternatives was no harder. We deployed several tools to achieve this goal; for instance, we remarked that instantiation of the paradigms with certain encryption fails to yield the required security. We further used the celebrated meta-reduction tool to show that the base encryption should satisfy strong security guarantees in order to provide secure constructions. Besides, we noted that concatenation of group elements destroys their algebraic structure and obstructs as a consequence verifiability.

Incremental constructions We took a step-wise and structured approach to progressively reinforce the security guarantees of our constructions. In fact, we started with outsider security in the two-user setting and finished with full insider multi-user security. We proposed on the way several add-ons for strengthening the security of the verification protocols. Such an approach helps understand the issues and choices involved in the constructions. It also offers flexible design options that accommodate to different settings.

Verifiability through group homomorphisms The keyword in our verification protocols is homomorphisms. In fact, whenever proving a property about some hidden information, the map underlying the property was homomorphic. This is manifest in the classes from which we instantiated the base primitives (signatures, (tag-based) encryption or commitments). With this choice, we were able to conduct efficiently the desired proofs through the so-called Σ protocols. We could also strengthen the security of the obtained protocols (negligible soundness error, concurrent zero-knowledge, and online non-transferability) at the expense of a tiny efficiency overhead. Finally, our protocols enjoy also the attractive feature of being convertible, at little extra cost and in *the standard model*, to non-interactive proofs to fit certain frameworks; this is definitely cheaper than resorting to the Groth-Sahai solution that is known for its expensive verification cost.

Notational Index

General

$\|a\|$	Absolute value of the real number a
$\max_A \mathrm{E}$	Maximum of the expression E when the variable A ranges over all the possible values
$\min_A \mathrm{E}$	Minimum of the expression E when the variable A ranges over all the possible values
$\log x$	Logarithm of x with respect to some unspecified base
$\ln n$	Logarithm of x in the base $e = \sum_{n=0}^{\infty} 1/n!$
$x \leftarrow y$	Assigning the value of y to x
$x \leftarrow \mathcal{A}$	Algorithm $\mathcal{A}$ returns the value x
$[a, b]$	Closed interval, i.e., the set of real numbers x in the range $a \leq x \leq b$
$[a, b)$	the set of real numbers x in the range $a \leq x < b$
$(a, b]$	the set of real numbers x in the range $a < x \leq b$

Bit Strings

ϵ	Empty string
$\perp$	Rejection symbol
$\overline{a}$	Bit complement of the string a
$a\|b$	Concatenation of the strings a and b
$\{0, 1\}^n$	Set of n-bit strings
$\{0, 1\}^*$	Set of all finite binary strings

Sets

$\emptyset$	Empty set
$\#A$	Cardinality of set A
$a \in A$	a is an element of set A
$a \notin A$	a is not an element of set A
$A \subset B$	Set A is properly contained in set B (i.e., A is not equal to B)
$A \subseteq B$	Set A is contained in or equal to set B
$A \cup B$	Union of sets A and B
$A \cap B$	Intersection of sets A and B
$A \backslash B$	Difference of sets A and B
$A \times B$	Cartesian product of sets A and B
$\mathbb{N}$	Set of natural numbers
$\mathbb{Z}$	Set of integers

L. El Aimani, *Verifiable Composition of Signature and Encryption*,
https://doi.org/10.1007/978-3-319-68112-2

$\mathbb{R}$	Set of real numbers
$\mathbb{Z}_N$	Set of integers modulo N (denoted also the set $\mathbb{Z}/N\mathbb{Z}$)
$\mathbb{Z}_N^\times$	Group of units in $\mathbb{Z}_N$

Groups

$(\mathbb{G}, +)$	Group $\mathbb{G}$ is denoted additively
$(\mathbb{G}, \cdot)$	Group $\mathbb{G}$ is denoted multiplicatively
$0_\mathbb{G}$	Identity in $(\mathbb{G}, +)$
$1_\mathbb{G}$	Identity in $(\mathbb{G}, \cdot)$
a^{-1}	Inverse of element a in a group denoted multiplicatively
$\langle g \rangle$	Group generated by the element g
$\mathsf{DL}_g(y)$	Discrete logarithm of group element y in base g

Functions

$f\colon A \rightarrow B$	f is function from set A to set B
$f\colon x \mapsto y$	Function f maps x to y
$x \mapsto y$	x is mapped to y (by some function)
$y \leftarrow f(x)$	f inputs x and returns y
$y \leftarrow f_x()$	f inputs x and returns y
$\perp$	Rejection symbol
$\perp \leftarrow f_x()$	f returns the rejection symbol on input x
f^{-1}	Inverse of bijective function f
poly	Polynomial function
negl	Negligible function, i.e., a function of order smaller than the inverse of any polynomial function

Integers

$a \text{ rem } b$	Remainder of the Euclidean division of a by b ($b \neq 0$)
$a \mid b$	a divides b
$\gcd(a, b)$	Greatest common divisor of integers a and b
$a \equiv b \bmod n$	a is congruent to b modulo n
$a^{-1} \bmod n$	Multiplicative inverse of a modulo n
$\Phi(n)$	Euler's totient function

Logic, Events, and Probabilities

$A \Rightarrow B$	A implies B (either A is false or B is true)
$b \leftarrow (c_1 = c_2)$	Bit b is assigned 1 if c_1 equals c_2, and 0 otherwise
$c_1 \stackrel{?}{=} c_2$	Logic expression whose value is 1 if c_1 equals c_2, and 0 otherwise
$\neg E$	Complement of event E
$E_1 \wedge E_2$	Intersection of event E_1 and event E_2
$E_1 \vee E_2$	Union of event E_1 and event E_2
$\Pr[E]$	Probability of event E
$\Pr[E_1 \mid E_2]$	Probability of event E_1 given event E_2
$\Pr[E_1, \ldots, E_n]$	Joint probability of events $E_1, \ldots, E_n$ (denoted also $\Pr[E_1 \wedge \ldots \wedge E_n]$)
$a \leftarrow D$	a is sampled from distribution D
$a \stackrel{R}{\leftarrow} S$	(Denoted also $a \in_R S$) a is selected uniformly at random from finite set S

Security Games

$x \leftarrow \mathcal{A}(1^\kappa)$	Algorithm $\mathcal{A}$ returns x on input security parameter κ
$\mathcal{I} \leftarrow \mathcal{A}$	Algorithm $\mathcal{A}$ returns state information $\mathcal{I}$

$x \xleftarrow{\tau \text{ operations}} \mathcal{A}()$	$\mathcal{A}$ returns x after performing τ elementary operations, that is, after running approximately in time τ
$\Gamma.\mathsf{cmp}$	Component cmp of the cryptographic system Γ; *cmp* can be an algorithm, a protocol, a key, etc.
$pk_{\mathsf{U}}/sk_{\mathsf{U}}$	Public/private key of user (entity) U
$\mathcal{A}^{\mathfrak{O}}$	Algorithm $\mathcal{A}$ has access to oracle $\mathfrak{O}$
$\mathfrak{O}: x \longmapsto f(x)$	Oracle $\mathfrak{O}$ returns $f(x)$ in response to query x
$\mathfrak{O}: x(\neq y) \longmapsto f(x)$	Oracle $\mathfrak{O}$ returns $f(x)$ in response to query x if it is different from y, otherwise $\mathfrak{O}$ returns the rejection symbol $\perp$
$\mathcal{A}^{\texttt{decrypt}^{\neg(c)}(sk,.)}$	$\mathcal{A}$ has access to decryption oracle `decrypt` which decrypts all ciphertexts except ciphertext c
$\mathcal{A}^{\texttt{decrypt}^{\neg(-,t)}(sk,.)}$	$\mathcal{A}$ has access to decryption oracle `decrypt` which decrypts w.r.t. all tags except tag t
$\mathcal{A}^{\texttt{decrypt}^{\neg(c,t)}(sk,.)}$	$\mathcal{A}$ has access to decryption oracle `decrypt` which accepts all queries (ciphertext, tag) except the pair formed by ciphertext c and tag t
$\mathcal{A}^{\texttt{decap}^{\neg(-,t)}(sk,.)}$	$\mathcal{A}$ has access to decapsulation oracle `decap` which decapsulates w.r.t. all tags except tag t
$c \leftarrow \mathcal{A}(\mathsf{find})$	$\mathcal{A}$ terminates the find stage, of the security game in question, by returning the challenge c
$r \leftarrow \mathcal{A}(\mathsf{guess})$	$\mathcal{A}$ terminates the guess stage, of the security game in question, by returning the response r
$\mathsf{Exp}_{\Gamma,\mathcal{A}}^{\mathsf{GOAL\text{-}ATK}}(\kappa)$	Security game where adversary $\mathcal{A}$ is challenged to break security notion GOAL-ATK of cryptosystem Γ on input security parameter κ
$\mathsf{Adv}_{\Gamma,\mathcal{A}}^{\mathsf{GOAL\text{-}ATK}}(\kappa)$	Advantage of adversary $\mathcal{A}$ in breaking security notion GOAL-ATK of cryptosystem Γ on input security parameter κ (this advantage is function of κ)

Proof Systems

$\pi_{\langle \mathsf{P}(w),\mathsf{V}\rangle}(x)$	Proof π is carried on instance x between prover P, with private input w, and verifier V
$\pi_w(x)$	Proof π is carried on instance x where w is the private input of the prover
$\langle done \mid out\rangle \leftarrow \pi$	Termination of proof π with the verifier either accepting ($out = 1$) or rejecting ($out = 0$)
`Cmt`	Algorithm used by the prover to compute the commitment in a Σ protocol
`Rsp`	Algorithm used in a Σ protocol to compute the response of the prover upon receipt of the challenge from the verifier
`Dcd`	Algorithm used by the verifier in a Σ protocol to compute the decision upon receipt of the prover's response
`trSim`(x)	Algorithm used to simulate the proof transcript of a Σ protocol
ZKP{S}	Zero-knowledge proof of the validity of statement S
ZKP$\{\mathsf{S}_1 \wedge \ldots \wedge \mathsf{S}_n\}$	Zero-knowledge proof of the validity of the joint statement $\mathsf{S}_1 \wedge \ldots \wedge \mathsf{S}_n$
ZKP$\{\mathsf{S}_1 \vee \ldots \vee \mathsf{S}_n\}$	Zero-knowledge proof of the validity of the disjunction $\mathsf{S}_1 \vee \ldots \vee \mathsf{S}_n$
ZKP$\{w\colon \mathsf{S}\}$	Zero-knowledge proof of the validity of statement S where w is the private witness of the prover
ZKPoK$\{w\colon \mathsf{S}\}$	Zero-knowledge proof of knowledge of w such that statement S (which involves w) holds true
$\xrightarrow{c}$	c is the message sent by the prover to the verifier
$\xleftarrow{b}$	b is the message sent by the verifier to the prover
$\xleftrightarrow{\text{ZKP\{S\}}}$	Prover issues the proof ZKP to prove the validity of statement S to the verifier

Index

A
adversary, 20
asymptotic security, 4
auxiliary string model, 27, 78

C
CDCS, 32
CDH, 18, 21
CHK transform, 126
CHK-like transform, 128, 134
chosen-ciphertext attack, 6
chosen-message attack, 4
chosen-plaintext attack, 6
commitment, 14, 15, 86, 93, 107, 118
  binding, 15
  hiding, 14
  injective, 15
common reference string, 27, 28, 100
completeness, 34, 42
concrete security, 5
confirmation protocol, 76, 99, 102
CtEaS, 85, 108
  indistinguishability, 108
  insider privacy, 131
CtEtS, 93, 100, 108
  invisibility, 96
  unforgeability, 94

D
Damgård-Pedersen, 60–62, 81
DDH, 18, 19, 21, 29, 61
DEM, 11, 68
  injective encryption, 11
denial protocol, 77, 99, 102
designated-verifier proofs, 36, 79
digital signature, 3, 49, 67, 85, 93, 100, 107, 110, 118
discrete logarithm, 15, 17, 18, 59, 78
disjunctive knowledge, 79

E
ElGamal, 21, 54, 75
EtS, 99, 107
  insider privacy, 132
  invisibility, 101
  security, 108
  unforgeability, 101
EtStE, 110
  indistinguishability, 111
  insider privacy, 133
  unforgeability, 111
EUF-CMA, 4, 37, 42
EUF-CMA-multi, 116

F
factoring, 16, 17
Fiat-Shamir, 28, 103
fully decryptable, 103
fully homomorphic, 53

G
GDH, 19, 21
Groth-Sahai, 29, 103

L. El Aimani, *Verifiable Composition of Signature and Encryption*,
https://doi.org/10.1007/978-3-319-68112-2

H

homomorphic, 73, 78, 98, 102
homomorphic encryption, 29, 53, 75, 80, 87, 103
homomorphic TBE, 122
hybrid encryption, 12, 21

I

IND-ATK, 8
IND-CCA, 43, 126
IND-CPA, 10
IND-PCA, 56, 62, 89, 108
IND-st-wCCA, 13, 119, 123, 126
indistinguishability, 8–10, 43
insider invisibility, 38
insider model, 33
insider privacy, 128
insider unforgeability, 37
interactive proof, 24
    completeness, 24
    soundness, 24
INV-CMA, 38
INV-CMA-multi, 117
INV-CPA, 9
INV-OT, 11
invisibility, 9, 11, 38

K

KEM, 10, 68
    tag-based, 13, 118
KEM/DEM paradigm, 12, 21, 67, 75, 110, 118, 130, 133
knowledge extractor, 25

M

meta-reduction, 23, 55–57, 61
multi-user CtEaS
    insider privacy, 135
multi-user CtEtS, 120, 121
multi-user EtS, 120, 121
    insider privacy, 135
multi-user invisibility, 117
multi-user security, 117
multi-user signcryption, 121
multi-user StE, 119
    insider privacy, 134
multi-user unforgeability, 116

N

negligible, 20
new StE, 68, 85, 108
    invisibility, 71
    strong unforgeability, 69
    unforgeability, 69
NIZK, 28, 29, 100, 103, 110
NM-CPA, 7, 56, 88
non-interactive proof, 28
non-malleable
    encryption, 7
    key generator, 57, 59, 89, 108
non-transferability, 35, 39, 42
    offline, 36
    online, 36, 79
noticeable, 20
NP, 26, 27, 50, 51, 100

O

one-time attack, 11
one-time signature, 5, 126, 128
one-wayness, 7
outsider invisibility, 38
outsider model, 33
overwhelming, 20
OW-CCA, 7, 55, 87
OW-CPA, 21

P

parallel composition, 85, 94, 131
Pedersen, 16, 98
PKI, 27, 78, 100
plaintext-checking attack, 6
proof of knowledge, 25, 26, 72, 78
    completeness, 25
    soundness, 25
public-coin protocol, 27, 77
public-key encryption, 6, 7, 12, 15, 21, 49, 85, 93, 100, 107, 110, 127

R

random oracle model (ROM), 22, 28, 103
reduction, 20
    advantage, 21
    arbitrary, 57, 61
    key-preserving, 21, 55, 56, 61, 87, 88
    useful, 57
registered key model, 29, 103
RSA, 17

S
selective-tag security, 13
sequential composition, 85
SEUF-CMA, 5, 37, 43
Sigma protocol, 27, 36, 79
signcryption, 39, 107, 132
SIND-CCA, 111
SINV-CMA, 70
soundness, 34, 42
soundness error, 25, 78
special soundness (SpS), 27
standard model, 22, 28
statistically indistinguishable, 68, 73
StE, 49, 67, 107
  indistinguishability, 108
  insider privacy, 130
  invisibility, 62
  unforgeability, 52
strong forgeability, 62, 67, 89, 93, 108, 129
strong indistinguishability, 111, 120
strong invisibility, 70
strong unforgeability, 5, 37, 43, 129, 130, 133, 134

T
tag-based encryption, 12, 51, 118, 122, 126, 127, 134
three-move protocol, 27, 28, 103
trusted authority, 28, 79, 99, 110

U
undeniable signature, 31, 60
unforgeability, 4, 37, 42

V
verifiability, 40, 109, 118, 127

W
WHPoK, 50

Z
zero-knowledge, 25, 26
  computational, 26, 77, 99, 102
  concurrent, 26, 27, 50, 77
  honest verifier (HVZK), 27, 77
  perfect, 26, 76, 99
  statistical, 26

Printed in the United States
By Bookmasters